G. B. Vine

M.A., Ph.D

Structural analysis

Longman London and New York

Longman Group Limited
Longman House
Burnt Mill, Harlow, Essex, UK

*Published in the United States of America
by Longman Inc., New York*

© **Longman Group Limited 1982**

First published 1982

British Library Cataloguing in Publication Data

Vine, G. B.
   Structural analysis. – (Longman technician
series: construction and civil engineering
sector)
   1. Structures, Theory of
I. Title
624.1′71   TA645   80–42209

   ISBN 0–582–41618–3

624

Printed in Singapore by
Selector Printing Co (Pte) Ltd

Longman Technician Series

Construction and Civil Engineering

General Editor — Construction and Civil Engineering

C. R. Bassett, B.Sc.

Formerly Principal Lecturer in the Department of Building and Surveying, Guildford County College of Technology

Books already published in this sector of the series:

Building organisations and procedures   *G. Forster*
Construction site studies — production, administration
    and personnel   *G. Forster*
Practical construction science   *B. J. Smith*
Construction science Volume 1   *B. J. Smith*
Construction science Volume 2   *B. J. Smith*
Construction mathematics Volume 1   *M. K. Jones*
Construction mathematics Volume 2   *M. K. Jones*
Construction surveying   *G. A. Scott*
Materials and structures   *R. Whitlow*
Construction technology Volume 1   *R. Chudley*
Construction technology Volume 2   *R. Chudley*
Construction technology Volume 3   *R. Chudley*
Construction technology Volume 4   *R. Chudley*
Maintenance and adaptation of buildings   *R. Chudley*
Building services and equipment Volume 1   *F. Hall*
Building services and equipment Volume 2   *F. Hall*
Building services and equipment Volume 3   *F. Hall*
Measurement Level 2 *M. Gardner*
Site surveying and levelling Level 2   *H. Rawlinson*

Contents

viii

Appendices 240

Acknowledgements

We are indebted to the following for permission to reproduce copyright materials:
The British Constructional Steelwork Association Ltd. for extracts from the *Structural Steelwork Handbook*; The British Standards Institution for our table 7.1 from *BS 449*.

Introduction

This book is set out in two main parts:

Part I :   General principles
Part II: The basic structures

in the belief that it is of great importance to grasp thoroughly the general principles which are few and simple before mastering the special approach required for each structure. It is no exaggeration to say that the general principles apply to any structure in every situation. Add to this the fact that the general principles are few and simple and it is clear that the easiest and quickest way to master the subject is to master the principles. To learn (without understanding the principles) all the multitude of ways to solve the different structures in their different situations is longer, harder, more tedious and probably plain boring compared with the effort required to be familiar with the few principles which apply in every situation. Inevitably some formulae need to be learnt. So – learn them!

There are many exercises to work at, so – have a go. Some are so easy you can write the answer straight down. Others need half an hour to solve. And a few are for the student who cannot solve too many of these problems. One thing is certain – structures can be interesting and the problems exciting to solve and I hope you find it so with at least some of the problems in this book.

Two further sections conclude the book:

Part III: Concrete
Part IV: The computer

Both these sections are meant as introductions. Firstly, an introduction to the special behaviour of reinforced concrete and prestressed concrete. Then, secondly, an introduction to programming the computer to solve problems in structural mechanics. You may have access to a computer terminal and for a small effort you can learn how to use it.

Part I

General principles

The basic principles are like the foundations of a structure – it's important to get them right!

Chapter 1

Structural analysis

1.1 The loading, the structure and mother earth

Structures have many purposes ranging from strength to beauty, but in this book we are concerned only with the ability of a structure to support a given set of loads with an adequate factor of safety and to carry those loads through the structure safely to the ground. Three elements interact therefore: the loading, the structure and mother earth. Our concern is to analyse the response of the structure sandwiched between the action of the loads and the reaction of the earth as shown in Fig. 1.1.

1.2 Important definitions and units

Definition 1.1 MASS is the amount of matter in a body.
It is measured in *kilograms* denoted by kg.

Corollary The mass of a body is the same wherever it is. The mass is independent of the gravitational forces and so a beam with a mass of 1000 kg on the earth still has a mass of 1000 kg on the moon even though it feels much lighter on the moon.

Definition 1.2 FORCE is what gives a body acceleration.
It is measured in *newtons* denoted by N and may be calculated from
force (N) = mass (kg) × acceleration (m/sec^2)

Corollary If a body is at rest then it has no net force acting upon it.

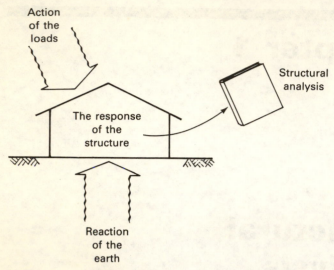

Fig.1.1

Definition 1.3 WEIGHT is the force of gravity acting on a body.

Corollary 1 weight (N) = mass (kg) × acceleration due to gravity (m/sec²)

Corollary 2 On earth the acceleration due to gravity is
g = 9.81 m/sec², so
weight (N) = mass (kg) × 9.81 (m/sec²)

Corollary 3 On the moon the acceleration due to gravity is
g = 1.64 m/sec², so
moon weight (N) = mass (kg) × 1.64 (m/sec²)

which explains why moon weights are much less than earth weights for the same masses.

Corollary 4 In outer space where there is no acceleration due to gravity we must have
outer-space weight = ZERO

Definition 1.4 The TONNE is 1000 kilograms.

Definition 1.5 The KILONEWTON is 1000 newtons.

These definitions should be learnt by heart and thoroughly understood. Most of the time (but not all the time) we shall be dealing with forces and not masses and hence the common until will be newtons (N) or kilonewtons (kN).

Definition 1.6 DENSITY is the mass per unit volume of a material. It is measured in kg/m³.

The densities of some common materials are:

water   1000 kg/m³
concrete   2400 kg/m³
steel   7200 kg/m³

1.3   Types of loading

The first and most obvious load that a structure must support is its own self weight and for large structures this is the major load. This is called the dead load of a structure.

The second type of load is called the live or imposed load. If it is not part of the dead load then it is the live load. For example, the cars on a bridge are live loads. Other live loads which the structure must be able to take arise from the forces of nature, for example wind or earthquake. It is not the prime purpose of the structure to resist these forces of nature but if the structure is to be useful it is obvious that it must do so (see Fig. 1.2a).

Let us now classify these loadings another way. They are shown diagrammatically in Fig. 1.2b–e. There is the point load, the uniformly distributed load (called UDL for short) and the uniformly varying load.

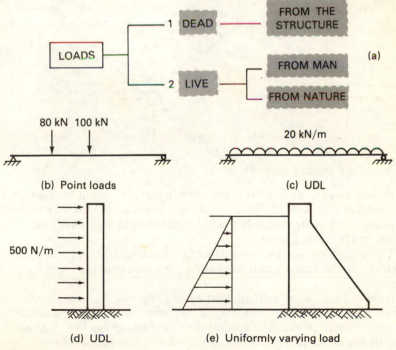

Fig. 1.2

8

1.4 Types of support

The supports are those parts of the structure which connect it to the ground. Since the ground affects the structure through the support it is important to distinguish the most important types. They are three: roller, pinned and fixed.

The simplest is the roller support shown in Figs. 1.3a, b, which allows both movement and rotation. The ground can only supply a reaction perpendicular to the line on which the support may roll.

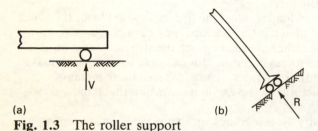

(a)   (b)

Fig. 1.3   The roller support

The next type is the pinned support which does not allow movement but does allow rotation. Examples are shown in Figs. 1.4a, b, and it can be seen that the support may transfer both a vertical and a horizontal thrust to the structure.

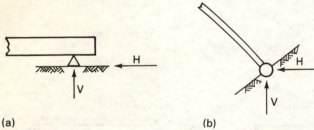

(a)   (b)

Fig. 1.4   The pinned support

Finally there is the fixed support allowing no movement or rotation. It is shown in Figs. 1.5a, b. Through this support the ground may supply a twisting effect to the structure (called a moment) in addition to the vertical and horizontal forces.

These supports deserve some attention because the behaviour of the structure and the analysis of it will vary a great deal with the type of support.

It should also be pointed out that by and large we are concerned with the forces that the ground gives to the structure and not the forces that the structure gives to the ground; so that the forces in Fig. 1.6a are correctly drawn while those of Fig. 1.6b are not. The support forces shown in Fig. 1.6b are the forces from the structure to the ground. If

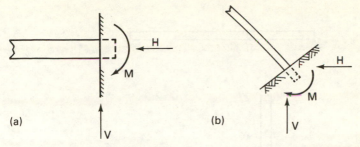

Fig. 1.5 The fixed support

you want to think about them both, then Fig. 1.6c is how it should be done. It is just a matter of distinguishing action and reaction – that is, the downward action of the structure on the ground and the upward reaction of the ground on the structure. Our concern is confined to the structure. It is the art of soil mechanics in which interest in the ground is uppermost.

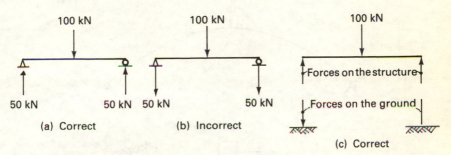

Fig. 1.6

1.5 Types of structure

There are very many different types of structure and doubtless a number of different ways of classifying them. In this book we have chosen six most commonly occurring structural forms. In this section we shall simply say what they are. Later on this chapter they will be looked at in a little detail but still in a general way, and then in Part II, Chapters 6 to 11, they will be analysed in detail. Here then are the six basic structures:

1 **the beam**   (*Ch. 6*)
2 **the column**   (*Ch. 7*)
3 **the truss**   (*Ch. 8*)
4 **the arch**   (*Ch. 9*)
5 **the wall**   (*Ch. 10*)
6 **the foundation**   (*Ch. 11*)

  These six structures are illustrated in Fig. 1.7 together with a typical loading.

10

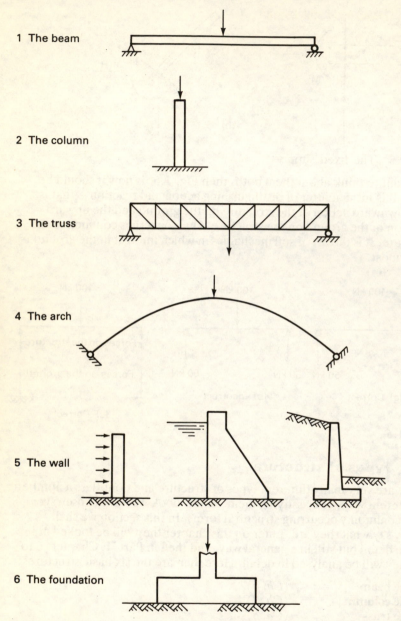

1 The beam

2 The column

3 The truss

4 The arch

5 The wall

6 The foundation

Fig. 1.7 The six most commonly occurring structural forms

1.6 Types of material

The variety of structural materials is very wide; nevertheless we may outline the main groups into which they fall. An obvious first choice is the metals, of which the most commonly used is steel. It is equally strong in tension or compression and its properties are reliable and well known.

A second choice in contrast to steel is a material like concrete. While it is strong in compression, it is very weak and unreliable in tension and, in addition to this, its properties can vary widely (compared with steel) even though the ingredients stay the same. Despite these drawbacks concrete still has a lot of advantages and when combined with steel to take the tensile forces it has proved itself to be a very versatile and valuable material. The analysis of steel and concrete working together is, however, a more complicated affair than taking them both separately and this is why concrete has a section on its own in this book. Other materials, like timber, may still be analysed by the same basic principles which apply to steel.

In this book the material itself has a second place and enters the problem with certain permissible stresses, elasticity, and so on. The first place is given to the analysis of the forces and stresses, which is largely independent of the material which goes to make up the structure. For example, a beam will have two support reactions equal to half the weight of the beam irrespective of whether the beam is made of steel, concrete or timber.

In short, the principles in this book will apply whatever material is being used . . . but the examples will be drawn from structures built from steel or timber or mass concrete or reinforced concrete.

1.7 The behaviour of beams and cantilevers

The beam spans between two supports and carries its load by deflecting and developing bending moments along its length. (See Fig. 1.8 and notice that the top edge of the beam is thrown into compression and the lower edge into tension.) One of the first aims in beam design is to ensure that these stresses in the upper and lower surfaces do not exceed the permissible stress for the material.

A little thought will show that these bending stresses are a maximum at midspan and a lot of thought may persuade you that they go down to zero at the support.

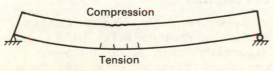

Fig. 1.8

12

In addition the beam experiences a varying shear force along its length going from a maximum at the support to zero at midspan. It is a rule in fact which we shall meet later that the maximum bending occurs where the shear is zero.

It is instructive to think about the internal forces in the beam at the midspan section. Referring to Fig. 1.9 we see the top half of the section is in compression and the lower half in tension. Halfway down the beam section there is a line on which there is no tension or compression and this is called the neutral axis.

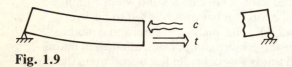

Fig. 1.9

The behaviour of the cantilever is similar to that of the beam except that the tensile zone of the cantilever is in the top and the maximum bending is at the support as shown in Fig. 1.10.

Beam and cantilever are often combined as in Fig. 1.11.

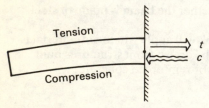

Fig. 1.10

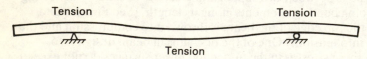

Fig. 1.11

1.8 The behaviour of columns

The main force that a column is designed to support is a load along its longitudinal axis called an axial load, as illustrated in Fig. 1.12a. Usually the loads are carried to the columns through the beams, as shown in Fig. 1.12b, in which the centre column still experiences an axial load of 500 kN.

However, Fig. 1.12b shows that the outer columns each have an eccentric loading of 250 kN (as opposed to an axial loading). The left-hand column is shown again in isolation in Fig. 1.12c, from which it is clear that the eccentric load gives a bending effect to the column as well as an axial force. Figure 1.12d shows the combined bending and axial effect of the eccentric load.

Two factors affect the strength of a column. The first is clearly the permissible stress for the material. The second factor is not at all obvious. If a column is long and slender then it may well fail due to buckling long before the permissible stresses have been reached. Failure at low stresses is quite possible and must be considered in the design of a column. Such low stress failure due to buckling is easily demonstrated: take a thin plastic ruler and see what load it can take as a column.

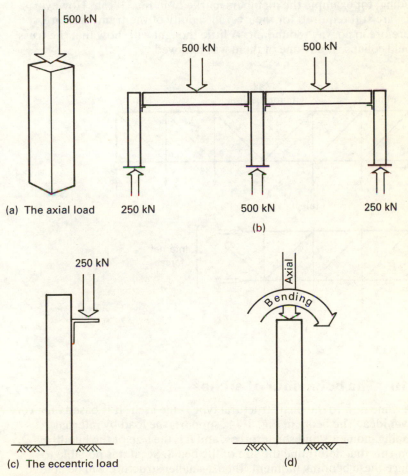

(a) The axial load

(b)

(c) The eccentric load

(d)

Fig. 1.12

1.9 The behaviour of trusses

The truss is a form of beam made up of a group of members pinned together at their ends. Each member is in simple tension or compression (although this is hardly realised in practice). Figure 1.13a shows a typical truss. As with the beam it is clear that the top members are in compression and the lower ones in tension. In the arrangement shown the diagonal members will be in tension (t) while the vertical members will be in compression, holding apart the upper and lower members.

Of course the maximum compression and tension will occur towards the centre.

As with the beam we may cut through the truss to show some of the internal forces, as shown in Fig. 1.13b.

Some of the members may have no force in them for a particular loading, for example the members marked x in Fig. 1.13a. However, they are still required for the overall rigidity of the frame and are therefore in no way redundant. A little thought will show that the truss would collapse if any one of them was removed.

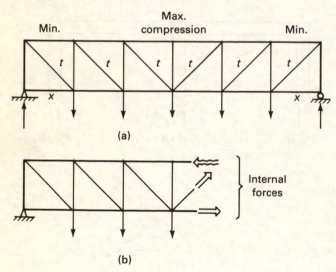

Fig. 1.13

1.10 The behaviour of arches

We come now to the next structural type – the arch. It is based on a very clever idea. The beam in Fig. 1.14a supports the load by internal bending moments and shear forces, and it is the size of the bending moments that determine the size of the beam. So, if it is possible to reduce these bending moments then a smaller structure will be able to carry the same load. This reduction in the bending of the beam is

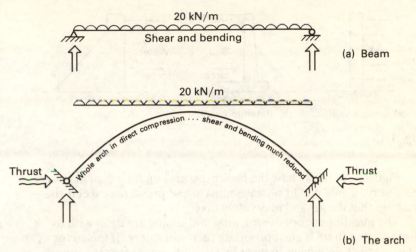

Fig. 1.14

brought about by arching it upwards and applying a horizontal thrust at
the supports, as shown in Fig. 1.14b. The loads now give their usual
downward bending effect but this is counteracted by the upward
bending effect of the thrust at the ends to result in a considerably
reduced bending of the arch section. Indeed, if the arch is parabolic and
the loading uniformly distributed, then the bending can be eliminated
completely. Of course, the arch still must experience something from
the loading and this is a direct compression along the arch shape. This is
the main force in the arch itself – the thrust (and not bending as in the
case of the beam). All this depends, of course, on being able to provide
the thrust at the supports. The supports at the abutments need to be
unyielding if they are to supply the thrust – if they yield then the thrust
goes down dramatically and bending will appear in the arch just as in a
beam.

1.11 The behaviour of walls

For the purpose of this section the term 'wall' includes retaining walls
and dams. The feature which distinguishes these structures from others
is that they have to resist horizontal loads, in contrast to most other
structures which support vertical loads. Figure 1.15 shows (a) a simple
wall with a wind loading, (b) a retaining wall with a horizontal loading
from some ground, and (c) a dam acted on by water.

Three main questions need consideration with a wall. They are:

1. *Are the pressures under the wall too great?*
2. *Is the wall near to overturning?*
3. *Is the wall near to sliding?*

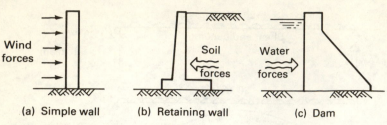

(a) Simple wall   (b) Retaining wall   (c) Dam

Fig. 1.15

Figure 1.16 shows how the base pressures under a dam may vary as the reservoir fills up. In no case must these pressures exceed the pressure that the ground may safely take.

The questions about overturning and sliding are dealt with by ensuring that there is an appropriate factor of safety. If the factor of safety against sliding has to be at least 2, this means that the forces available to stop the wall from sliding, for example the friction under the base of the wall, must be at least twice the forces causing sliding. Usually this is the difficult factor of safety to obtain as sliding is the critical condition in most cases.

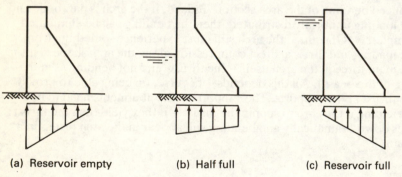

(a) Reservoir empty   (b) Half full   (c) Reservoir full

Fig. 1.16

1.12 The behaviour of foundations

A foundation transmits the forces in the structure to the ground itself. Foundations are necessary because the stresses in the material of the structure are far higher than any stresses that the ground can withstand. For example, a certain concrete column may be made of concrete some 50 times stronger than the ground on which it rests. As a result the stresses in the column when under load may be some 50 times greater than the permissible ground pressure and so the column would be pushed easily into the soil as soon as the loads came down the column.

In essence, therefore, the foundation spreads the load from a column over a sufficiently large area for the ground to be able to take it. These points are illustrated in Fig. 1.17.

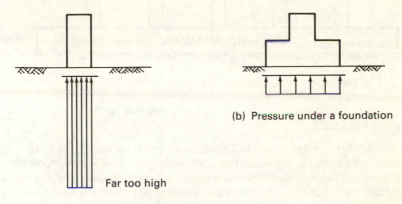

(b) Pressure under a foundation

Far too high

(a) Pressure directly under a column

Fig. 1.17

The simplest type of foundation is the column base. If the column only carries an axial load then clearly there will be a uniform pressure distribution under it, as in Fig. 1.18a. If the column has some bending in it, however, the pressure will increase on one side and decrease on the other, as illustrated in Fig. 1.18b. Further increase in the bending may produce uplift, as shown in Fig. 1.18c, but this is undesirable and should be avoided.

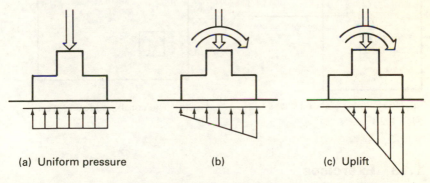

(a) Uniform pressure   (b)   (c) Uplift

Fig. 1.18

Sometimes if there are a number of columns in a line, a strip foundation may be used (Fig. 1.19). Or again, there may be some columns too close to each other for the required isolated bases to fit in. In Fig. 1.20a the base for column A overlaps the base for column B. In

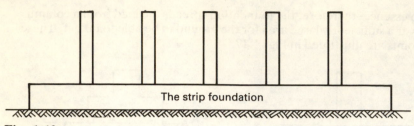

Fig. 1.19

such a case a combined foundation for both columns could be used as shown in Fig. 1.20b.

This section brings us round full circle to the section beginning this chapter – the loading, the structure and mother earth. We have looked at the types of loading and six basic structural forms which will carry these loads. We have pointed out that the structure lies between the action of the loads and the reaction of the earth. This last section has been concerned with the action of the structure on the earth and it is the foundation which is the link between them. The foundation is there to carry the structure and to carry it safely to the earth.

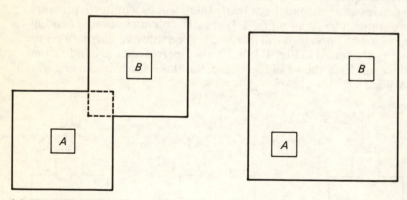

(a) No room for isolated bases

(b) The combined foundation

Fig. 1.20

1.13 Exercises

1 Write down the different ways in which the loads on a structure can be classified.
2 Define mass, force and weight.
3 Distinguish clearly between mass and weight.
4 Compare the weight of a mass of 1 tonne on the earth with its weight on the moon. Express the weights in kN.

5 Name the three main types of support.

6 Name the six structures with which this book is concerned.

7 Choose one of the six structural forms and explain as simply as possible its purpose and how it supports its loads. Your account must be about 100 words and written for someone who has no knowledge of the subject. When you have finished give it to someone to see if they can understand it.

8 Make a list of the places where you have seen examples of the six structures.

9 In which part of a beam are there tensile forces?

10 In which part of a cantilever are there compressive forces?

11 Is it possible for a column to have tension in it when a downward load acts on it?

12 With what type of joint are the members of a truss connected together?

13 Is it possible for a member of a truss to experience bending?

14 There is a connection between the number of joints (j) and the number of members (m) in a triangulated truss. Draw out roughly a few trusses and find values of j and m for each of them. Can you guess the relationship between j and m?

16 Explain the effect on an arch if one of the abutments yields (a) vertically, and (b) horizontally.

17 What are the three main factors that have to be considered when designing a retaining wall?

18 Sketch a retaining wall and mark on it the point about which it will rotate if it overturns.

19 Name the three main types of foundation.

20 Explain why it is necessary for a structure to have a foundation and why the structure cannot rest directly on the ground.

Chapter 2

Equilibrium

2.1 The meaning of equilibrium

In this chapter the basic equations for equilibrium are going to be set out. They will be few and simple but it is absolutely essential to understand them.

Definition 2.1
Equilibrium is a state of rest in which all the forces acting on a structure balance each other.

Corollary If a structure is in equilibrium then for any given direction the forces trying to move the structure one way will be balanced by the forces trying to move the structure the opposite way, and for any given point the forces trying to rotate the structure clockwise will be balanced by the forces trying to rotate the structure anticlockwise.

These ideas need now to be made more precise before they can be put to good use.

2.2 Force, moment and component

The idea of a force has already been spoken of in section 1.2. A force has magnitude and direction, e.g. 6 kN *up*, 100 kN *down*, 57 kN *horizontally to the right*, and so on. A force also has a turning effect about a point. Look at Fig. 2.1a. The turning effect of the 12 kN force about the point A is 12 × 3 = 36 kN m *clockwise*. This is called a bending moment or simply a moment. Similarly, in Fig. 2.lb, the moment of the 15 kN force about A is 15 × 4 = 60 kN m *anticlockwise*. Referring to Fig. 2.1c, what is the moment of the 10 kN force . . . about A? . . . about B?

Referring to Fig. 2.1d, we may now give the following definition of the moment of a force:

Definition 2.2

The moment of a force about a point	= force ×	the perpendicular distance from the point to the force

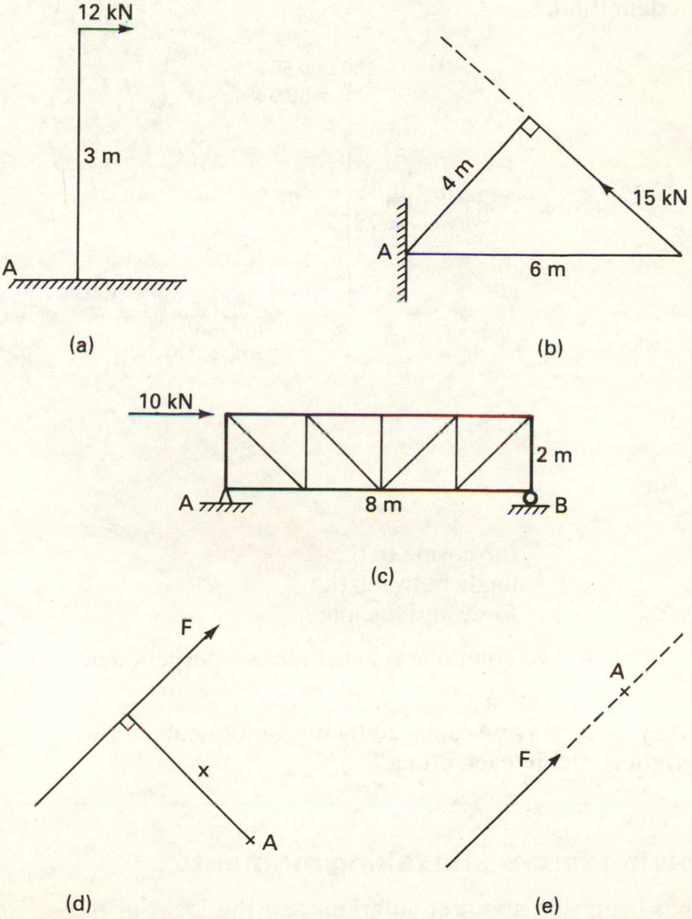

Fig. 2.1

22

A special case occurs when the force passes through the point, as in Fig. 2.1e. In this case the force has no moment about the point. Hence we have:

Corollary If a force passes through a point then it has no moment about that point.

Now we turn to the value of a force in a direction other than the one in which it is acting. Look at Fig. 2.2a. What is the 100 kN worth horizontally? Clearly it is something rather less than 100 kN. It is in fact 100 cos 30° = 86.6 kN. This is called *the horizontal component* of the force. Similarly, *the vertical component* is 100 cos 60° = 50.0 kN. Any force may be replaced by two components at right angles, and so Fig. 2.2b shows the two components of the 100 kN force and they are completely equivalent to it. This is of fundamental importance so we have our next definition.

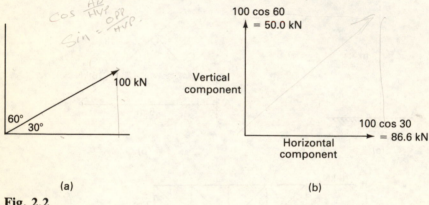

(a)   (b)

Fig. 2.2

Definition 2.3

The component of a force along a line = force × the cosine of the angle between the force and the line

Corollary 1 A force has no component in a direction perpendicular to its line of action.

Corollary 2 Any force may be replaced by its components in two directions at right angles to each other.

2.3 Resolving forces and taking moments

From what has been said about equilibrium and the ideas of the component of a force and the moment of a force, we may now state the two basic principles which must apply to the forces acting on a structure which is in equilibrium. They should be learnt by heart and thoroughly understood.

Basic Principle 1
For any given direction,

The sum of the
force components =
one way

The sum of the
force components
the opposite way

Basic Principle 2
For any given point,

The sum of the
clockwise =
moments

The sum of the
anticlockwise
moments

    Applying these two conditions is called *resolving forces in a given direction* and *taking moments about a given point*. To understand this is to grasp the essential idea at the heart of structural analysis.

    Before turning to see how these principles are used a further word is necessary to explain how many times we may usefully apply them. The short answer is, three times. This may be explained in the following way:

Equilibrium $\Rightarrow$ The structure is at rest
$\Rightarrow$ 1. no movement parallel to x axis
2. no movement parallel to y axis
3. no rotation
$\Rightarrow$ there are 3 conditions to be fulfilled.

    These three conditions amount to resolving forces twice and taking moments once. This is the general case for a plane structure. You will see in some of the examples that follow that only two conditions are applied. This is when there are no horizontal forces on the structure and so it seems unnecessary to apply the third condition of resolving horizontally.

Example 1
Calculate the reactions for the beam in Fig. 2.3.
Taking moments about B,

   clockwise moments = anticlockwise moments

Hence

   $R_1 \times 12 = 40 \times 10 + 50 \times 8 + 60 \times 6 + 70 \times 4 + 80 \times 2$

Therefore,

   $R_1 = 133.3$ kN

24

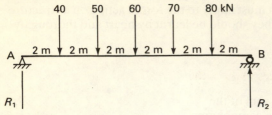

Fig. 2.3

Resolving vertically,

    force components up = force components down

Hence,

    $R_1 + R_2 = 40 + 50 + 60 + 70 + 80$

Therefore,

    $R_2 = 166.7$ kN

Example 2

Calculate the reactions of the truss in Fig. 2.4.

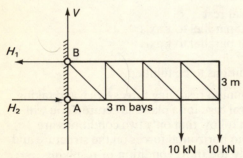

Fig. 2.4

Resolving vertically,

    forces up = forces down

So $V = 20$ kN

Taking moments about A,

    anticlockwise moments = clockwise moments

Therefore

    $H_1 \times 3 = 10 \times 9 + 10 \times 12$

Hence,

$H_1 = 70.0$ kN

Resolving horizontally,

forces to the left = forces to the right

So $H_1 = H_2$

Hence,

$H_2 = 70.0$ kN

Example 3

Figure 2.5 shows a cantilever. The support reaction consists of a moment M and a vertical reaction V. Calculate M and V.

Firstly we resolve vertically:

forces up = forces down

So, $V = 12$ kN

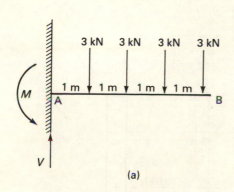

(a)

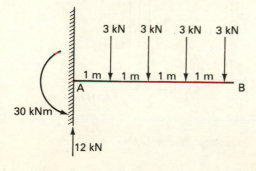

(b)

Fig. 2.5

26

Secondly, we take moments about A:

anticlockwise moments = clockwise moments

So, $M = 3 \times 1 + 3 \times 2 + 3 \times 3 + 3 \times 4$

or $M = 30$ kN m

The result is shown in Fig. 2.5b.

Example 4

A wall is shown in Fig. 2.6. A wind force acts on its side as well as its own self weight. Determine the reactions at the base of the wall.

We simply resolve or take moments in order to obtain the unknowns in the simplest possible way.

Resolve vertically:

forces up = forces down

So $R = 72$ kN

Resolve horizontally:

forces pushing to the left = forces pushing to the right

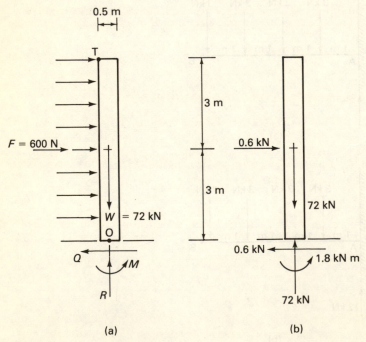

(a)   (b)

Fig. 2.6

Hence,

$Q = 600 \text{ N} = 0.6 \text{ kN}$

Taking moments about the centre of the base, O,

anticlockwise moments = clockwise moments

Therefore,

$M = 600 \times 3 = 1800 \text{ N m} = 1.80 \text{ kN m}$

And so we obtain the final result shown in Fig. 2.6b.

Can it really be this simple? Yes! – if you choose the right directions in which to resolve the forces (horizontal and vertical are the most popular) and if you choose the right points about which to take moments (a support or the centre of a section is usually the best). You may take moments about the moon if you wish – the equations will be harder to solve but they will still give the same answers.

For example, if we had resolved at 30° and 40° to the horizontal and taken moments about T (Fig. 2.6a) we would have obtained

$$R \cos 60 + 0.6 \cos 30 = 72 \cos 60 + Q \cos 30$$
and $$R \cos 50 + 0.6 \cos 40 = 72 \cos 50 + Q \cos 40$$
and $$M + R \times 0.25 + 0.6 \times 3 = 72 \times 0.25 + Q \times 6.0$$

The equations are true but not very helpful. Who wants to solve equations like this? Nevertheless, their solution is as before,

$R = 72 \text{ kN}, Q = 0.6 \text{ kN and } M = 1.8 \text{ kN m}$.

2.4 The resultant and the equilibrant

We begun this section with two definitions and a corollary.

Definition 2.4
The EQUILIBRANT of a group of forces is that single force which will hold the group in equilibrium.

Definition 2.5
The RESULTANT of a group of forces is that single force which has the same effect as the whole group.

Corollary The equilibrant and resultant are equal and opposite to each other.

Example 5
Look at the forces in Fig. 2.7a. They are all equal and equally spaced. A little common sense is all that is required here to show that the resultant must be 60 kN acting at the centre of the forces as shown. The

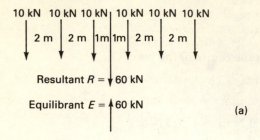

10 kN 10 kN 10 kN 10 kN 10 kN 10 kN

2 m   2 m   1m 1m   2 m   2 m

Resultant R = 60 kN

Equilibrant E = 60 kN

(a)

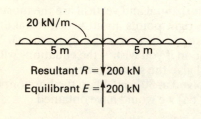

20 kN/m

5 m   5 m

Resultant R = 200 kN

Equilibrant E = 200 kN

(b)

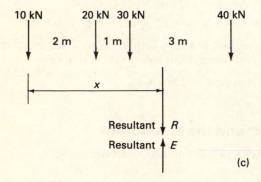

10 kN   20 kN 30 kN   40 kN

2 m   1 m   3 m

x

Resultant R

Resultant E

(c)

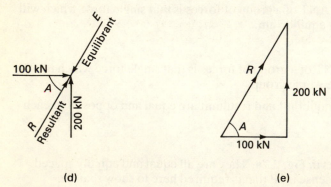

100 kN

A

R Resultant

Equilibrant E

200 kN

(d)

R

200 kN

A

100 kN

(e)

Fig. 2.7

equilibrant is exactly opposite to the resultant. In a similar way, a uniformly distributed load such as that shown in Fig. 27b has a resultant and equilibrant acting at its centre of value $20 \times 10 = 200$ kN.

Example 6

Look now at Fig. 2.7c. Common sense says that the equilibrant must be parallel to the given forces but what is its magnitude and where does it act? We follow the two principles of equilibrium.

Resolving vertically,

$E = 10 + 20 + 30 + 40 = 100$ kN

Taking moments about the left-hand force,

$E \times x = 20 \times 2 + 30 \times 3 + 40 \times 6$

Hence,

$x = 3.7$ m

The resultant is the same as E but in the opposite direction.

Example 7

Figure 2.7d shows two forces of 100 kN and 200 kN at right angles. Calculate the magnitude and direction of the resultant.

The shortest way to solve this is to look at the triangle of forces shown in Fig. 2.7e.

Clearly, $\quad \tan A = \dfrac{200}{100} \quad$ So, $\quad A = 63.4°$

By Pythagoras' theorem,

$R = \sqrt{100^2 + 200^2} \quad$ So, $\quad R = 223.6$ kN

The equilibrant is the same as the resultant but the other way.
Alternatively we may resolve vertically and horizontally.
Remembering that $\cos(90 - A) = \sin A$ we obtain

$R \sin A = 200$
$R \cos A = 100$

Dividing one by the other gives, $\tan A = 2$
Hence,

$A = 63.4°$

Using the fact that $\sin^2 A + \cos^2 A = 1$ we may square and add the equations to give $R^2 = 200^2 + 100^2$

So $\quad R = 223.6$ kN.

2.5　The centre of gravity

We begin with a definition:

Definition　2.6
The CENTRE OF GRAVITY of a set of weights is the point through which their resultant acts.

It is worth obtaining a formula which will give the position of the centre of gravity. Look at Fig. 2.8. The set of loads W_1, W_2, W_3, W_4 have a resultant R which acts a distance $\bar{x}$ from some point O. Let us try to find a general expression for $\bar{x}$.

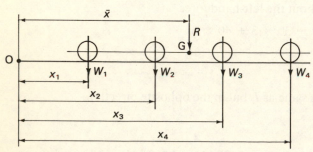

Fig. 2.8

Resolving vertically,

$$R = W_1 + W_2 + W_3 + W_4 = \Sigma W$$

Taking moments about O,

$$R \times \bar{x} = W_1 x_1 + W_2 x_2 + W_3 x_3 + W_4 x_4 = \Sigma W x$$

Thus we have:

Formula　2.1
$$\bar{x} = \frac{\Sigma W x}{\Sigma W}$$

This is well worth committing to memory. The symbol Σ simple means 'the sum of'.

Example 8
Determine the position of the centre of gravity of the loads shown in Fig. 2.9.

We have

$$\bar{x} = \frac{\Sigma W x}{\Sigma W}$$

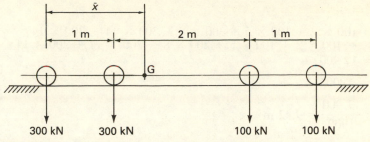

Fig. 2.9

So, measuring all distances to the left-hand load, we obtain

$$\bar{x} = \frac{300 \times 0 + 300 \times 1 + 100 \times 3 + 100 \times 4}{300 + 300 + 100 + 100}$$

Hence,

$$\bar{x} = 1.25 \text{ m}$$

Example 9

A group of columns is shown in plan in Fig.2.10. Locate the centre of gravity G.

In this case we have two dimensions to calculate but the principle is the same.

Firstly,

$$\bar{x} = \frac{\Sigma Wx}{\Sigma W}$$

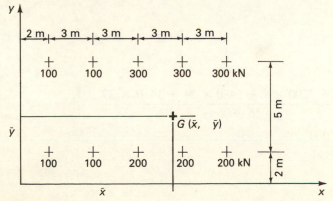

Fig. 2.10

Putting

$$\Sigma Wx = 100 \times 2 + 100 \times 5 + 300 \times 8 + 300 \times 11 + 300 \times 14$$
$$+ 100 \times 2 + 100 \times 5 + 200 \times 8 + 200 \times 11 + 200 \times 14$$
$$= 17\ 900$$

and $\Sigma W = 1900$

gives $\bar{x} = \dfrac{17\ 900}{1900} = 9.42$ m

Secondly,

$$\bar{y} = \frac{\Sigma Wy}{\Sigma W}$$

where

$$\Sigma Wy = 100 \times 7 + 100 \times 7 + 300 \times 7 + 300 \times 7 + 300 \times 7$$
$$+ 100 \times 2 + 100 \times 2 + 200 \times 2 + 200 \times 2 + 200 \times 2$$
$$= 9300$$

Hence,

$$\bar{y} = \frac{9300}{1900} = 4.89 \text{ m}$$

Example 10

The cross-section of a concrete dam is shown in Fig.2.11a and it is required to find the position of the centre of gravity. Taking 1 m length of the dam, the weights of the various parts and their distances to the front edge are given in Fig. 2.11b

From

$$\bar{x} = \frac{\Sigma Wx}{\Sigma W}$$

we obtain

$$\bar{x} = \frac{960 \times 2 + 7200 \times 9 + 6480 \times 24 + 1440 \times 29}{16\ 080}$$

So $\quad \bar{x} = \dfrac{264\ 000}{16\ 080}$

and finally,

$$\bar{x} = 16.42 \text{ m}$$

33

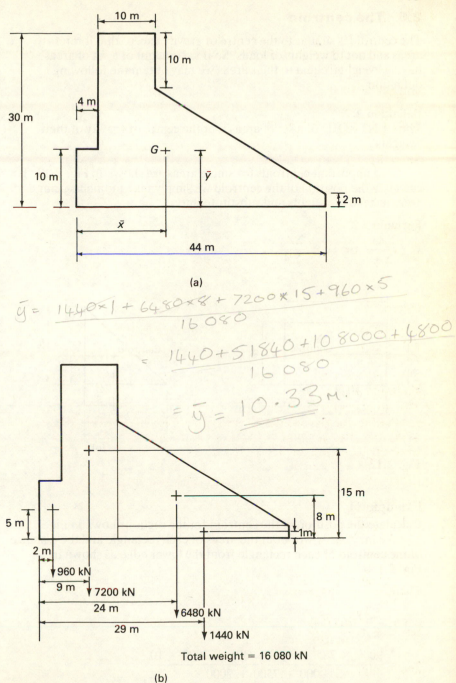

(a)

$$\bar{y} = \frac{1440 \times 1 + 6480 \times 8 + 7200 \times 15 + 960 \times 5}{16080}$$

$$= \frac{1440 + 51840 + 108000 + 4800}{16080}$$

$$= \bar{y} = 10.33 \text{ M}.$$

Total weight = 16 080 kN

(b)

Fig. 2.11

2.6 The centroid

The centroid is similar to the centre of gravity except that it refers to areas and not to weights or loads. So if we thought of a set of areas having 'weights' equal to their areas we may obtain the following definition.

Definition 2.7
The CENTROID of a set of areas is at the centre of gravity of their 'weights'.

Two important centroids for single areas are shown in Fig. 2.12. To calculate the position of the centroid we simply take moments using areas instead of weights and substitute into:

Formula 2.2

$$\bar{x} = \frac{\Sigma Ax}{\Sigma A} \quad \text{or} \quad y = \frac{\Sigma Ay}{\Sigma A}$$

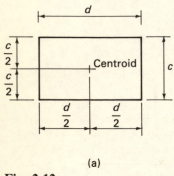

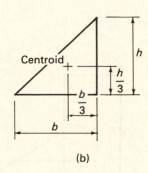

(a)   (b)

Fig. 2.12

Example 11
Calculate the position of the centroid for the section shown in Fig.2.13a.
We may easily work out the areas of each rectangle and the distance of the centroid of each rectangle from the lower edge as shown in Fig. 2.13b

Then,

$$\bar{y} = \frac{\Sigma Ay}{\Sigma A}$$

$$= \frac{9000 \times 785 + 7500 \times 395 + 2000 \times 10}{9000 + 7500 + 2000}$$

$$= 543 \text{ mm}$$

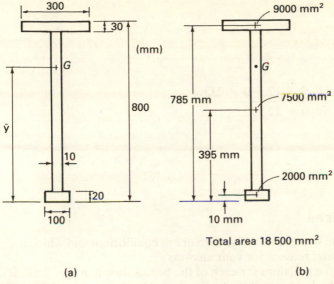

Total area 18 500 mm²

(a)   (b)

Fig. 2.13

Example 12

A section is built up from a Universal Beam (UB) and a plate as shown in Fig. 2.14a. Locate the centroid.

The important point to grasp here is that the UB may be dealt with as a single area. From the tables we may find the area of the UB to be 125.3 cm² or 12 530 mm². Its overall depth is 467.4 mm so its own centroid is half of this amount above the lower edge. These dimensions and those for the plate are given in Fig. 2.14b.

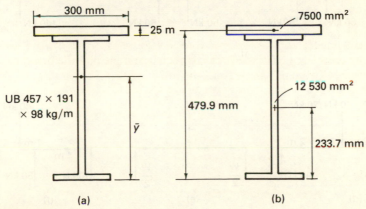

(a)   (b)

Fig. 2.14

36

From

$$\bar{y} = \frac{\Sigma A y}{\Sigma A}$$

we have

$$\bar{y} = \frac{7500 \times 479.9 + 12\ 530 \times 233.7}{7500 + 12\ 530}$$

So,

$$\bar{y} = 325.9 \text{ mm}$$

2.7 Exercises

1 Which of the structures in Fig. 2.15 are in equilibrium and which are not? Give brief reasons for your answers.
2 Determine the reactions for each of the beams shown in Fig. 2.16. If the loading includes a uniformly distributed load then the total value of the distributed load may be considered as acting at the centre of its length.
3 Calculate the reactions for each cantilever shown in Fig. 2.17. Note that the reactions consist of two forces and a moment.
4 Some columns are shown in Fig. 2.18. Calculate the reactions.

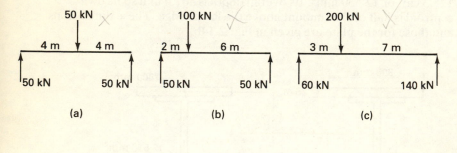

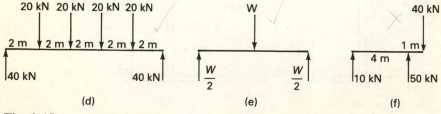

Fig. 2.15

37

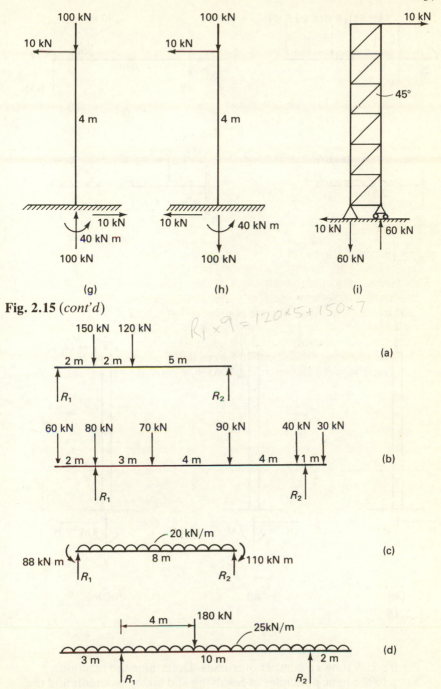

(g) (h) (i)

Fig. 2.15 (*cont'd*)

$R_1 \times 9 = 120 \times 5 + 150 \times 7$

150 kN 120 kN

2 m 2 m 5 m

(a)

R_1 R_2

60 kN 80 kN 70 kN 90 kN 40 kN 30 kN

2 m 3 m 4 m 4 m 1 m

(b)

R_1 R_2

20 kN/m

88 kN m 8 m 110 kN m

(c)

R_1 R_2

4 m 180 kN

25kN/m

3 m 10 m 2 m

(d)

R_1 R_2

Fig. 2.16

38

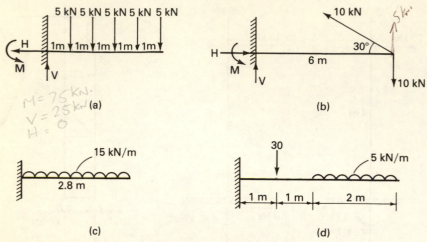

M = 75 kN·
V = 25 kN
H = 0

(a)

(b)

(c)

(d)

Fig. 2.17

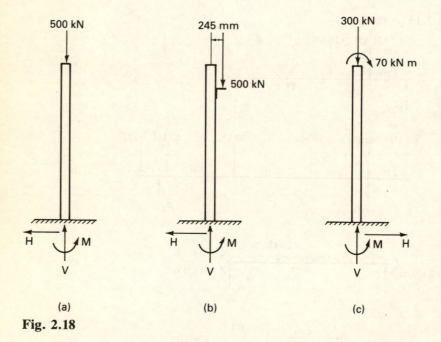

(a)

(b)

(c)

Fig. 2.18

5 Figure 2.19 shows a number of trusses. Determine the reactions. Keep to the basic principles of resolving and taking moments and the answers will be easy to obtain.

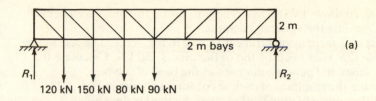

(a)

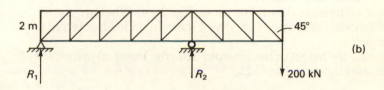

(b)

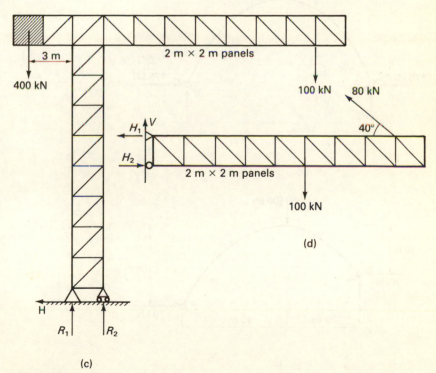

(d)

(c)

Fig. 2.19

40

6 Figure 2.20 shows two arches. The horizontal thrust is given in each
case. Calculate the vertical reactions.
7 The pier of a suspension bridge is shown in Fig. 2.21. The tension in
the cable that runs over the top of the pier is 850 kN. Calculate the
thrust, shear and bending moment at the base of the pier.
8 Determine the resultant of each set of forces shown in Fig. 2.22.
9 Locate the centre of gravity of each of the load trains shown in
Fig. 2.23.
10 Six columns lie on a straight line at 4 m centres. Their loads are 512,
625, 596, 577, 601, 611 kN. How far is the centre of gravity of these
loads from the first column?
11 A set of columns is shown in plan in Fig. 2.24. What are the
coordinates of the centre of gravity of the column loads referred to
the origin?
12 Determine the height of the centroid from the lower edge for each of
the sections shown in Fig. 2.25.

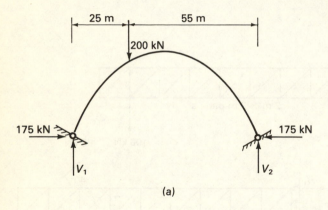

(a)

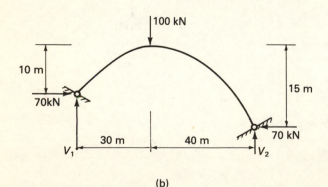

(b)

Fig. 2.20

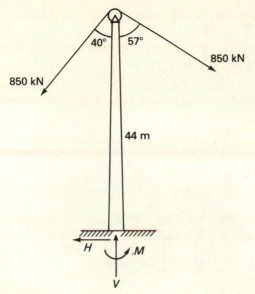

Fig. 2.21

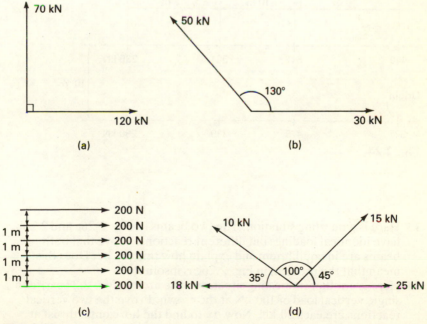

Fig. 2.22

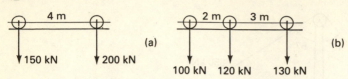

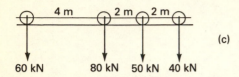

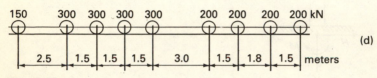

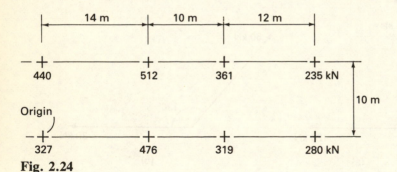

Fig. 2.23

Fig. 2.24

13 Here is a puzzling situation. The two beams in Fig. 2.26a and 2.26b have identical loadings but different reactions. Prove that both beams are in equilibrium and explain how this can be. Does this mean that the problem in Fig. 2.26c is insoluble?

14 Here is another interesting situation. The arch in Fig. 2.27 carries a single vertical load of 100 kN at the crown. Prove the two vertical reactions are each 50 kN. Now try to find the horizontal thrust at each abutment. Why is it impossible to determine?

43

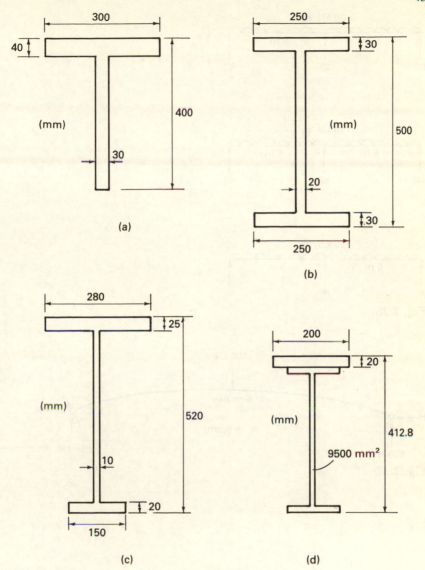

Fig. 2.25

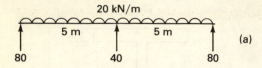

20 kN/m

5 m   5 m   (a)

80   40   80

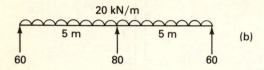

20 kN/m

5 m   5 m   (b)

60   80   60

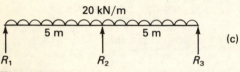

20 kN/m

5 m   5 m   (c)

R_1 R_2 R_3

Fig. 2.26

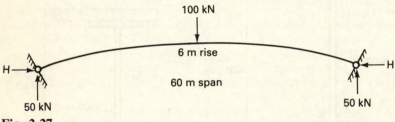

100 kN

6 m rise

60 m span

H →   ← H

50 kN   50 kN

Fig. 2.27

Chapter 3

Internal forces and moments

3.1 The idea of internal forces

A structure may be acted upon by a number of external forces but we must ask – are these the only forces that the structure experiences? A little thought will show that the external forces give rise to a set of internal forces. Indeed the external loads are transferred through the structure to the supports through the internal forces in a continuous flow of action and reaction.

Another way of looking at the internal forces is to consider two halves of the structure, the left half and the right half, for example. Now, what holds these two parts together? Clearly some internal forces must act between the two halves. Look at the bar in simple tension in Fig. 3.1a. What are the internal forces at X? These are shown in Fig. 3.1b where the bar is shown cut through at X. The force that the

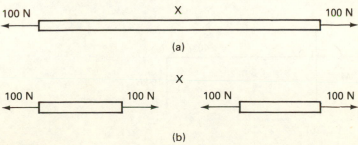

(a)

(b)

Fig. 3.1

right half exerts on the left is like an external force to the left half, and of course this force is the same as the force exerted by the left half on the right half.

3.2 Cutting the structure into two parts

We are now led to consider a cut through the structure in order to reveal the internal forces at a particular point. This idea is extremely important. The action of one part on the other may be reduced in general to two forces and a moment acting at the centroid of the section. Of the two forces, one acts perpendicular to the section and is called the normal force, N, and the other acts in the plane of the section and is called the shear force, Q.

Example 1

Figure 3.2a shows a cantilever. What are the internal forces at X?

We cut the cantilever through at X and consider just one half, in this case the right half. The left half must have exerted a normal force (N), shear force (Q) and moment (M) on the right half, and these are shown in Fig. 3.2b. The basic principles of equilibrium – resolving and taking moments – will easily yield N, Q and M.

Resolve horizontally, $N = 10 \cos 40$ So, $N = 7.66$ kN
Resolve vertically, $Q + 10 \cos 50 = 10$ So, $Q = 3.57$ kN

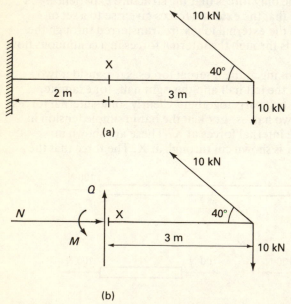

Fig. 3.2

Take moments about X (using the horizontal and vertical components of the inclined force),

$$M + 10 \cos 50 \times 3 = 10 \times 3 \qquad \text{So, } M = 23.1 \text{ kN m}$$

Example 2

A portal frame is shown in Fig. 3.3a with the reactions at the foot of each column already calculated (by methods beyond the scope of this book). Calculate the internal forces at X.

We cut the structure at X – we will take the cut through the beam – and consider the left half of the frame as in Fig. 3.3b. Resolving horizontally,

forces to left = forces to right
So $\quad N = 30$ kN

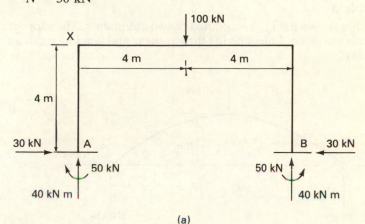

(a)

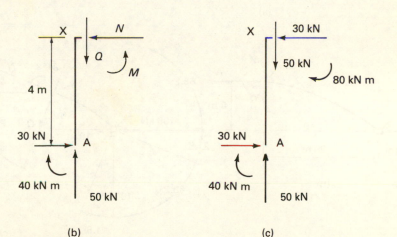

(b)   (c)

Fig. 3.3

48

Resolving vertically,

forces down = forces up
So, $Q = 50$ kN

Taking moments about X,

  anticlockwise moments = clockwise moments

Hence, $M + 30 \times 4 = 40$

So $M = -80$ kN m

The minus sign means that the direction chosen for M in Fig. 3.3b was not the actual direction of the moment. Changing the direction of M gives the result in Fig. 3.3c

Example 3

An arch is shown in Fig. 3.4a pinned to two abutments. The support reactions are given. Calculate the thrust, shear and bending moment at the crown.

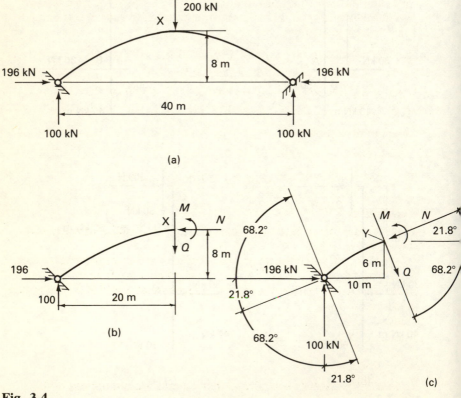

(a)

(b)

(c)

Fig. 3.4

Cut the structure at X just to the left of the central load to give Fig. 3.4b.

Resolving horizontally,

$N = 196$ kN

Resolving vertically,

$Q = 100$ kN

Taking moments about X,

$M + 196 \times 8 = 100 \times 20$

So $M = 432$ kN m

Example 4

Calculate the internal forces at the left quarter point for the arch of Fig. 3.4a. The slope of the arch has been calculated to be 21.8° to the horizontal.

Figure 3.4c shows the arch cut at the quarter point. With a litttle care even this more complicated problem is easily solved. We resolve parallel to N and Q (not horizontally and vertically unless you like complicated equations).

Resolve parallel to N: $N = 196 \cos 21.8 + 100 \cos 68.2$

So $N = 219.1$ kN

Resolve parallel to Q: $Q + 196 \cos 68.2 = 100 \cos 21.8$

So $Q = 20.1$ kN

Take moments about Y:

$M + 196 \times 6 = 100 \times 10$

So $M = -176$ kN m (that is 176 kN m hogging)

Example 5

What are the forces in the members marked X, Y and Z of the truss shown in Fig. 3.5a?

In this case when the structure is cut we have three internal forces and no moment, as shown in Fig. 3.5b. To find X we take moments about the point P where Y and Z intersect so that Y and Z do not appear in the equations.
Hence,

$X \times 2 + 10 \times 2 = 25 \times 4$

So, $X = 40.0$ kN

To find Y we resolve vertically so that X and Z do not appear in the equations.

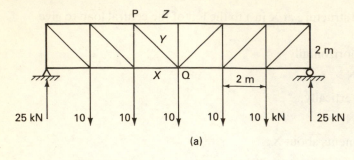

(a)

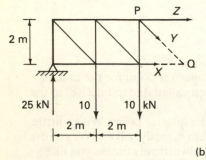

(b)

Fig. 3.5

Hence,

$$Y \cos 45 + 10 + 10 = 25 \qquad \text{So,} \quad Y = 7.1 \text{ kN}$$

And finally, for Z we take moments about the point Q where X and Y intersect so that neither X nor Y appear in the equations. Hence,

$$Z \times 2 + 25 \times 6 = 10 \times 4 + 10 \times 2$$

So $Z = -45.0$ kN (the minus sign showing Z is in compression)

Sufficient examples have now been done to show that the two principles of equilibrium are enough to give the internal forces in a structure. In general, the reactions have to be found first (the cantilever is the exception to this) and then the structure is cut at the point at which the internal forces are required. Only one half of the structure is considered and resolving or taking moments are applied to this half.

3.3   Some questions and answers

Q: *Does it matter which directions I choose for* N, Q *and* M?
A: No. If you get it wrong the answer will be negative. But if you make
   M a sagging moment then you have a helpful sign convention of
   sagging positive, hogging negative. (See Fig. 3.6 for the usual
   conventions.)

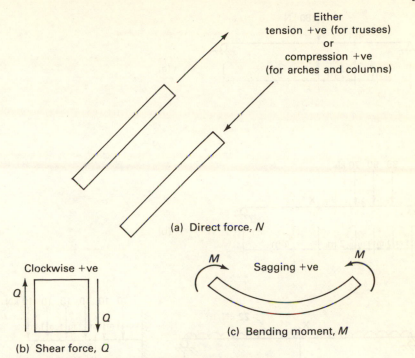

Either
tension +ve (for trusses)
or
compression +ve
(for arches and columns)

(a) Direct force, N

Clockwise +ve

Sagging +ve

Q

Q

(c) Bending moment, M

(b) Shear force, Q

Fig. 3.6 Sign conventions for internal forces

Q: Does it matter in which direction I resolve?
A: No – but there is a best direction. In general resolve parallel to N and parallel to Q. The exception is the truss.

Q: About which point should I take moments?
A: Always take moments about the point in which you cut the structure (that is, the centroid of the cut section). The only exception is the case of the truss.

Q: What is the general approach for the truss?
A: The cut will always be through 3 members – say X, Y and Z. If you want X take moments about the point where Y and Z intersect. If Y and Z do not intersect because they are parallel then just resolve perpendicular to them.

Q: Finally, does it matter which half of the structure I choose to work on?
A: No – but choose the half which looks the simpler of the two.

3.4 Exercises

1 Find the internal thrust, shear and bending moment at the point X in each of the structures shown in Fig. 3.7

52

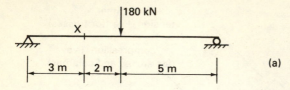

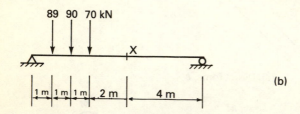

180 kN

X

3 m | 2 m | 5 m

(a)

89 90 70 kN

X

1 m | 1 m | 1 m | 2 m | 4 m

(b)

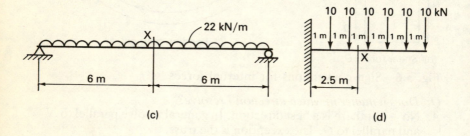

22 kN/m

X

6 m | 6 m

(c)

10 10 10 10 10 10 kN

1 m | 1 m | 1 m | 1 m | 1 m | 1 m

X

2.5 m

(d)

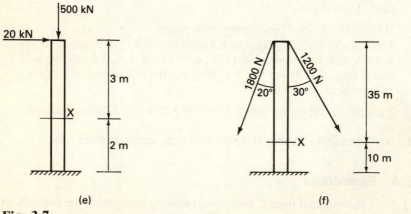

500 kN

20 kN

3 m

X

2 m

(e)

1800 N   1200 N

20° 30°

X

35 m

10 m

(f)

Fig. 3.7

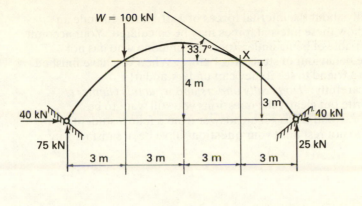

(g)

Fig. 3.7 (*cont.*)

2 Determine the forces in the members marked X, Y and Z in the trusses shown in Fig. 3.8

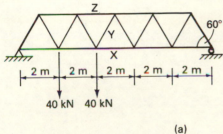

(a)

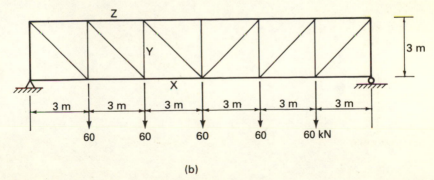

(b)

Fig. 3.8

3 Write briefly about the internal forces of a structure. Include an outline of how these internal forces may be calculated. Your account must be capable of being understood by someone who has not studied the behaviour of structures before. When you have finished it give it to a friend to see if they can understand it!

4 Consider carefully: *'Do I really understand what this chapter is about?'* Write out clearly the questions you still want to be answered . . . and try to get an answer from a fellow student or a lecturer. Do not rest until your questions have been satisfactorily answered.

Chapter 4

Section properties

4.1 The importance of this chapter

This chapter forms the bridge between force and stress. Equilibrium is concerned chiefly with forces but the ability of a structure to take certain forces has to be decided in the light of the stresses in the material. The force on the structure is one thing; the stress in the material of the structure is another thing altogether. The strength of the material is given in terms of a stress – the permissible stress – and not a force. It is meaningless to say 'the permissible force in steel is 165 N' because the amount of force that steel can take depends on how much steel there is. However, to say the permissible stress in steel is 165 N/mm² is to say something very valuable about the strength of steel. Now, it is the next chapter which deals with stress but before we can come to it we need to do some mathematics. The significance of what we work out will become plain in the next chapter on stress. Meanwhile take it on trust that this is an important chapter forming the vital link between force and stress as illustrated in Fig. 4.1.

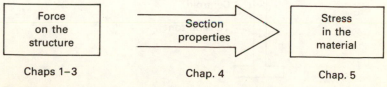

Fig. 4.1

4.2 The area of a section

Area is a well-known idea and needs no introduction here. A few words of warning about units are necessary, however. Area is measured in square metres (m^2) or square centimetres (cm^2) or square millimetres (mm^2) and it is important to be able to convert quickly from one to the other. Generally speaking the best unit to work with for structural sections is mm^2. For areas on the outside of a structure (for example the base of a foundation) it is best to use m^2. In the Universal Beam and Column tables (Appendices 1 and 2) the areas are given in cm^2 and it will be necessary to change these units to mm^2.

Conversion factors to remember
$1\ cm^2 = 10^2\ mm^2$
$1\ m^2 = 10^6\ mm^2$

4.3 The centroid

The formula for the position of the centroid has already been given in Chapter 2 but we quote it again here for completeness.

Centroid formulae to remember

$$\bar{x} = \frac{\Sigma Ax}{\Sigma A}$$

$$\bar{y} = \frac{\Sigma Ay}{\Sigma A}$$

The centroid of a section is a very important point because any bending of the section takes place about the axis through the centroid called the neutral axis (NA) (Fig. 4.2).

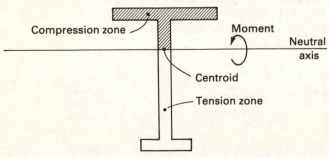

Fig. 4.2

57

4.4 The second moment of area

At this point we break fresh ground. One of the most important properties of a section is its second moment of area denoted by I. It is this property which will give the stresses due to a bending moment. The proof is given in Appendix 3 in order to keep this part as short as possible.

Definition 4.1
The SECOND MOMENT OF AREA of a small area A about an axis distant r from it is $I = Ar^2$ (see Fig. 4.3a). The units are m^4 or cm^4 or mm^4.

Example 1
Calculate the second moment of area of a rectangle 50 mm × 100 mm.

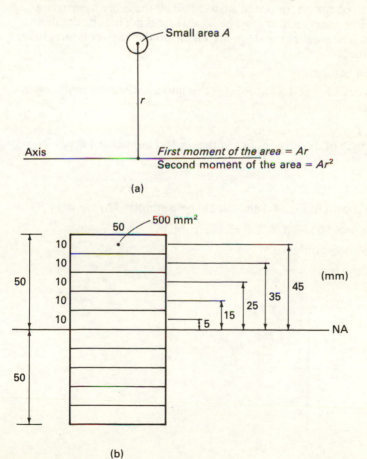

(a)

(b)

Fig. 4.3

58

Let us split the rectangle into strips 10 mm wide as shown in Fig. 4.3b. Summing each area times the square of its distance to the neutral axis we find:

$$I = 500 \times 45^2 + 500 \times 35^2 + 500 \times 25^2 + 500 \times 15^2 + 500 \times 5^2$$
$$+ 500 \times 45^2 + 500 \times 35^2 + 500 \times 25^2 + 500\ 053\ 15^2 + 500 \times 5^2$$

$$= 4\ 125\ 000\ \text{mm}^4$$

However, if smaller strips of 5 mm width had been taken the result would have been

$$I = 4\ 156\ 250\ \text{mm}^4$$

and with 1 mm strips,

$$I = 4\ 166\ 250\ \text{mm}^4.$$

There is, of course, only one answer and all these are approximations to it. The exact answer must be based on taking infinitesimally small strips. The smaller the strips are taken, so the answer homes in on the true value,

$$I = 4\ 166\ 666\ \text{mm}^4$$

Now we really need a formula for this and avoiding the mathematics we obtain,

Formula 4.1

For a rectangle of dimensions $b \times d$ the second moment of area is,

$$I = \frac{bd^3}{12}$$

This is shown in Fig. 4.4 and should be committed to memory.

If we put $b = 50$ and $d = 100$ we then obtain

$$I = 4\ 166\ 666\ \text{mm}^4$$

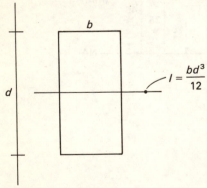

Fig. 4.4

which we should round up to,

$I = 4.17 \times 10^6$ mm^4.

Now we must turn to a very important theorem which must be thoroughly learnt and understood. Look at Fig. 4.5. An area has a second moment of area about an axis through its centroid of I_g. What is its second moment of area, I, about a parallel axis a distance h away? The following theorem gives the result.

The parallel axis theorem
$I = I_g + Ah^2$ (see Fig. 4.5)

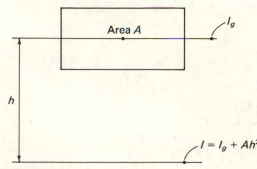

Fig. 4.5

Example 2
Calculate the value of I for the section shown in Fig. 4.6.

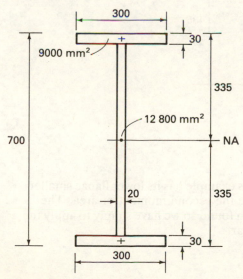

Fig. 4.6

60

We simply use the parallel axis theorem for each part and then add them up together.
Hence,

$$I = \frac{300 \times 30^3}{12} + 9000 \times 335^2 \quad \text{(top flange)}$$

$$+ \frac{20 \times 640^3}{12} + 12\,800 \times 0^2 \quad \text{(web)}$$

$$+ \frac{300 \times 30^3}{12} + 9000 \times 335^2 \quad \text{(bottom flange)}$$

So $I = 2458 \times 10^6 \text{ mm}^4$

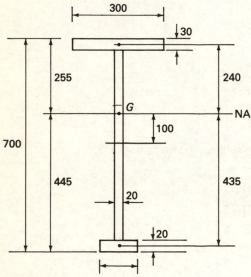

Fig. 4.7

Example 3 (Fig. 4.7)

A similar section to the previous example has its lower flange smaller than its upper flange. Calculate the second moment of area. The position of the centroid has been found so we have simply to apply the parallel axis theorem for each part.

From

$$I = I_g + Ah^2$$

we obtain

$$I = \frac{300 \times 30^3}{12} + 9000 \times 240^2 \qquad \text{(top flange)}$$

$$+ \frac{20 \times 650^3}{12} + 13\,000 \times 100^2 \qquad \text{(web)}$$

$$+ \frac{100 \times 20^3}{12} + 2000 \times 435^2 \qquad \text{(lower flange)}$$

So $\quad I = 1485 \times 10^6 \text{ mm}^4$

Look again at Fig. 4.7. The dimensions on the left of the section (255, 455 mm) fix the position of the centroid. The dimensions on the right (240, 435, 100 mm) are the distances from the centroids of each element (the flanges and the web) to the centroid of the whole section, and these distances are the values of h to be used in the parallel axis theorem.

Example 4 (Fig. 4.8)

A Universal Beam (UB457 $\times$ 191 $\times$ 98 kg/m) has a steel plate 300 mm $\times$ 30 mm welded to its top flange. Determine the second moment of area of the section. The relevant properties of the UB are: depth $D = 467.4$ mm, area $A = 125.3$ cm^2 and second moment of area $I = 45\,717$ cm^4 as taken from the second tables.

We simply treat the UB as a whole. The position of the centroid is given by,

$$\bar{y} = \frac{\Sigma Ay}{\Sigma A} = \frac{9000 \times 482.4 + 125.3 \times 10^2 \times 233.7}{9000 + 125.3 \times 10^2}$$

$$= 337.7 \text{ mm}$$

Hence we may obtain the dimensions in Fig. 4.8b.

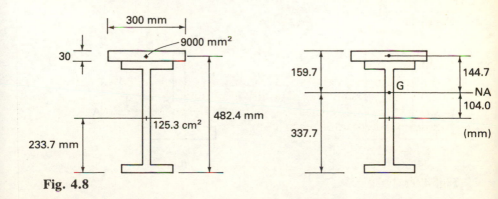

Fig. 4.8

Using the parallel axis theorem gives

$$I = \frac{300 \times 30^3}{12} + 9000 \times 144.7^2 \quad \text{(plate)}$$

$$+ \; 45\,717 \times 10^4 + 125.3 \times 10^2 \times 104.0^2 \quad \text{(UB)}$$

$$= 782 \times 10^6 \; \text{mm}^4$$

4.5 The section modulus

Definition 4.2
The SECTION MODULUS of a section is

$$Z = \frac{I}{y_{max}}$$

where I is the second moment of area and y_{max} is the greatest distance from the NA to an extreme edge (Figs. 4.9a, b). The units are m^3 or cm^3 or mm^3.

Corollary For a rectangle of dimensions $b \times d$ the section modulus is

$$Z = \frac{bd^2}{6}$$

Can you see why?

So Z is simple to calculate once we have the value of I. First we find which edge is furthest from the neutral axis and how far away it is and this is then y_{max}. Then we calculate I/y_{max} and this is the value of the section modulus, Z. Its importance is simple. When a section is subjected to a bending moment it is the edge which is furthest from the neutral axis which has the greatest stress in it and by using Z and the bending moment M we can calculate the stress.

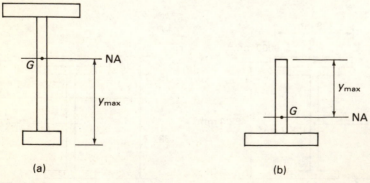

(a)   (b)

Fig. 4.9

4.6 The radius of gyration

We come to a section property which is not as popular as the second
moment of area but it does have its uses (chiefly to do with the buckling
of slender columns), so we include it here. It does have some practical
meaning as can be seen from Fig. 4.10. Suppose a certain section has an
area A and second moment of area I. At what distance from the neutral
axis should one place the whole area so that the second moment of area
is unchanged? The answer is to place the area (half on one side of the
NA and half on the other) at a distance equal to the radius of gyration.
Using Definition 4.1 then gives $I = Ar^2$ which, on rearrangement gives
the following formula.

Formula 4.2

Radius of gyration, $r = \sqrt{\dfrac{I}{A}}$

Corollary For a rectangle $I = bd^3/12$, $A = bd$

So $\qquad r = \dfrac{d}{\sqrt{12}} \doteq \dfrac{d}{3.5}$

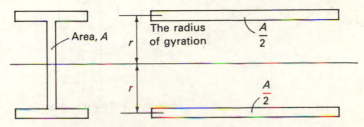

Each section has the same value of I

(a)   (b)

Fig. 4.10

4.7 The kern points

The kern points are both important and interesting. A vertical load
placed at the centre of a column will clearly throw the whole column
into compression. However, if the load is moved away from the centre
then the compression will be reduced along the edge furthest from the
load. At a certain point (the kern point) the compression will be
reduced to zero and if the load moves any further then tension will
appear. This is shown in Figs. 4.11a and 4.11b. For a rectangle the kern
points lie on the edges of the middle third (Fig. 4.11c). The shape traced
out by all the kern points gives the 'core' of a section (Figs. 4.11d, e). A
longitudinal load within the core will not cause tension but a load
outside the core will give rise to tension.

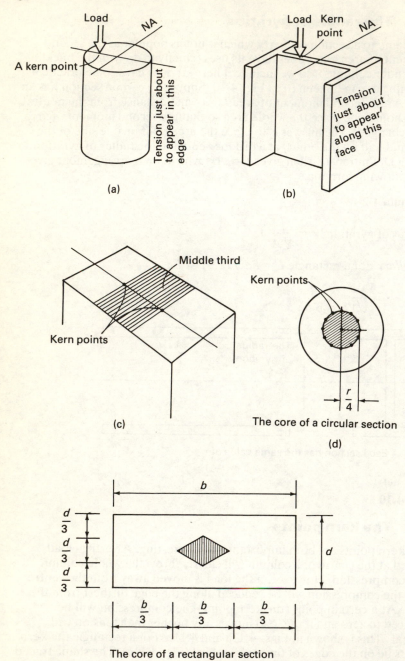

Fig. 4.11

The importance of this is seen in arch design where the aim is to ensure that the thrust in the arch rib remains within the core of the rib. Again with the design of a concrete gravity dam, it is important that the resultant force on any section passes within the core of the section. Let us put these results together in a form which they can be used.

Figure 4.12 shows a section with the kern points at distances k_1 and k_2 above and below the neutral axis respectively. Suppose the upper and lower edges are y_1 and y_2 from the neutral axis and the radius of gyration (referred to the given neutral axis) is r, then we obtain the following formula for the kern point positions.

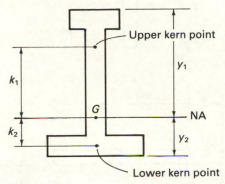

Fig. 4.12

Formula 4.3

$$k_1 = \frac{r^2}{y_2} \quad \text{(upper kern point)}$$

and
$$k_2 = \frac{r^2}{y_1} \quad \text{(lower kern point)}$$

Corollary For a rectangle $k_1 = k_2 = d/6$, which means the kern points lie on the edges of the middle third of the section (see Fig. 4.11c.)

Example 5 (Fig. 4.13a)
Determine all the section properties given in this chapter for the given section.

The area

Clearly, $A = 200 \times 40 + 30 \times 360$ mm^2

Expressing this in a number of different ways, we have

$A = 18\,800$ mm$^2 = 18.8 \times 10^3$ mm^2
$\quad = 188$ cm^2
$\quad = 0.0188$ m$^2 = 18.8 \times 10^{-3}$ m^2

66

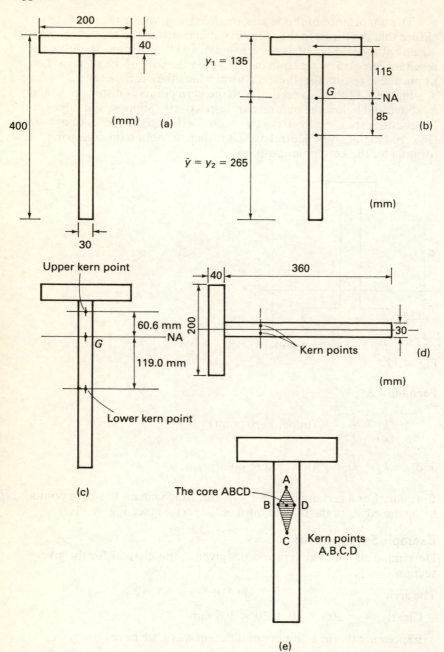

Fig. 4.13

The centroid

From $\qquad \bar{y} = \dfrac{\Sigma Ay}{\Sigma A}$

we have $\qquad \bar{y} = \dfrac{(40 \times 200) \times 380 + (30 \times 360) \times 180}{18\ 800}$

$$= 265 \text{ mm}$$

from which are obtained the dimensions in Fig. 4.13b.

The second moment of area

Using the parallel axis theorem,

$$I = I_g + Ah^2$$

we have, $\qquad I = \dfrac{200 \times 40^3}{12} + 8000 \times 115^2 \qquad \text{(flange)}$

$$+ \dfrac{30 \times 360^3}{12} + 10\ 800 \times 85^2 \qquad \text{(web)}$$

$$= 302 \times 10^6 \text{ mm}^4$$

The section modulus

We know $\qquad Z = I/y_{max}$

so $\qquad Z = \dfrac{302 \times 10^6}{265}$

$$= 1.14 \times 10^6 \text{ mm}^3$$

which of course refers to the lower edge.

The radius of gyration

From $\qquad r = \sqrt{\dfrac{I}{A}}$

we obtain $\qquad r = \sqrt{\dfrac{302 \times 10^6}{18.8 \times 10^3}} = 127 \text{ mm}$

The kern points

The upper and lower kern points are respectively,

$$k_1 = \dfrac{r^2}{y_2} \quad \text{and} \quad k_2 = \dfrac{r^2}{y_1} \quad \text{from the NA.}$$

Hence,

$$k_1 = 60.6 \text{ mm} \quad \text{and} \quad k_2 = 119.0 \text{ mm}$$

and these are shown in Fig. 4.13c.

The core

    To find the core it is necessary to find the kern points lying on the other axis. Turning the section on its side for convenience as in Fig. 4.13d we can easily obtain I, A, r and the kern points associated with the given axis.

So $$I = \frac{40 \times 200^3}{12} + \frac{360 \times 30^3}{12}$$

Hence $$I = 27.5 \times 10^6 \text{ mm}^4$$

So $$r = \sqrt{\frac{27.5 \times 10^6}{18.8 \times 10^3}} = 38.2 \text{ mm}$$

Clearly the kern points are equidistant, k, from the neutral axis, where

$$k = \frac{r^2}{y} = \frac{38.2^2}{100} = 14.6 \text{ mm}$$

So these kern points lie just within the web. Joining these kern points to those found previously gives the core of the section (Fig. 4.13e).

4.8 Formulae to remember

Here are summarized the formulae given in this chapter:

Centroid

$$\bar{x} = \frac{\Sigma Ax}{\Sigma A} \qquad \bar{y} = \frac{\Sigma Ay}{\Sigma A}$$

Second moment of area

parallel axis theorem $\qquad I = I_g + Ah^2$

rectangle $\qquad I = \dfrac{bd^3}{12}$

Section modulus

$$Z = \frac{I}{y_{max}}$$

rectangle $\qquad Z = \dfrac{bd^2}{6}$

Radius of gyration

$$r = \sqrt{\frac{I}{A}}$$

Kern points

$$k_1 = \frac{r^2}{y_2} \qquad k_2 = \frac{r^2}{y_1}$$

4.9 Exercises

1 Name the six section properties with which this chapter is concerned. Quote the formula for each one from memory.
2 Determine the area, the second moment of area and the section modulus for each of the rectangles shown in Fig. 4.14 for bending

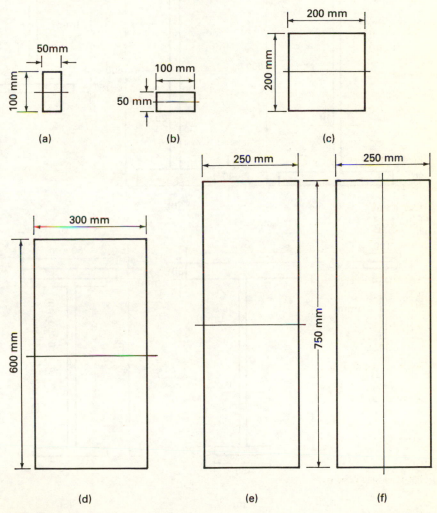

Fig. 4.14

70

about the given axis. Why is rectangle (a) different from (b)? What is the significance of the areas of (e) and (f) being the same, but their second moments of area being different?

3 Determine the second moment of area and the section modulus for each of the symmetrical sections shown in Fig. 4.15.

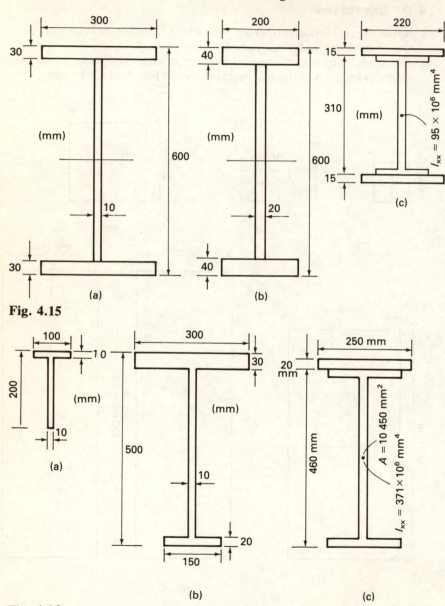

Fig. 4.15

Fig. 4.16

4 Determine the second moment of area and the section modulus for each of the unsymmetrical sections shown in Fig. 4.16.
5 Locate the kern points for the sections in Fig. 4.17.
6 A beam section is shown in Fig. 4.18. Determine the position of the core of the section and draw a scale drawing of it.
7 A slab foundation has the shape given in Fig. 4.19. Locate the core and draw a scale plan of it within the shape of the foundation.

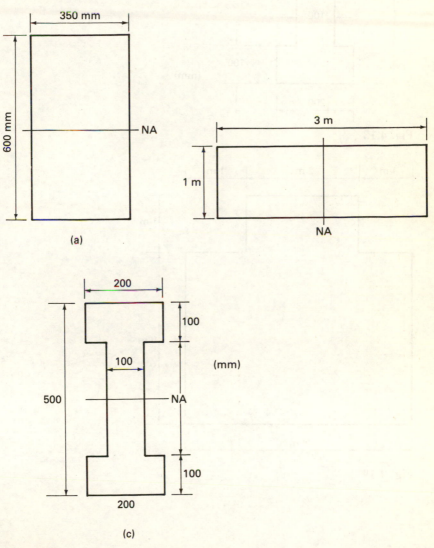

Fig. 4.17

72

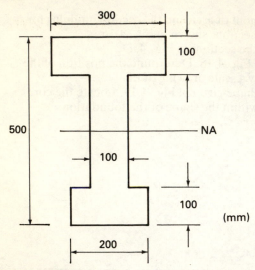

Fig. 4.18

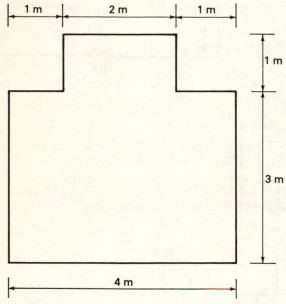

Fig. 4.19

Chapter 5

Stress

5.1 The definition of stress

We begin first with a definition.

Definition 5.1
STRESS is force per unit area. Its units are N/m^2 or more usually N/mm^2.

Note: The words '*stress*' and '*pressure*' are very similar in meaning. They both have the same definition (force per unit area) and the same units. Nevertheless they are generally used in different contexts. Stress refers to what is happening on the *inside* of a material while pressure refers to what is happening on the *outside*. So we might say,

the stress *in* the concrete is 8 N/mm^2

or, the stress *in* the steel is 165 N/mm^2

but, the wind pressure *on* the wall is 100 N/m^2

or, the water pressure *on* the dam is 500 kN/m^2

or, the pressure *on* the ground is 200 kN/m^2

or, the pressure *under* the foundation is 200 kN/m^2.

Corollary If a force acts on an area then the average stress over the area may be found from

$$\text{Stress} = \frac{\text{force}}{\text{area}}$$

However, life is not always that simple. You will notice that we have said 'average stress' and that is because the stress may not be evenly distributed over the area, and the way in which the stress is distributed depends on the type of force being exerted on the area. In Chapter 3 we looked at the internal forces in a structural member and saw that they were three in number – the normal force, N, the shear force, Q, and a bending moment, M. Each of these three – N, Q and M – give rise to different distributions of stress and we shall look at each in turn in the following sections. Nevertheless there really are only two types of stress corresponding exactly to the two different types of force. As there are normal forces and shear forces so there are also normal stresses and shear stresses.

Definition 5.2
The NORMAL STRESSES on a section are those stresses acting perpendicular to the section.

Corollary Normal stresses can arise not only from a normal force but also from a bending moment.

Definition 5.3
The SHEAR STRESSES on a section are those stresses acting in the plane of the section.

Corollary Shear stresses can arise not only from a shear force but also from a torsional moment.

These points are illustrated in Figs. 5.1 and 5.2. In the next sections we shall give the formulae which will enable the stresses to be calculated from N, Q and M.

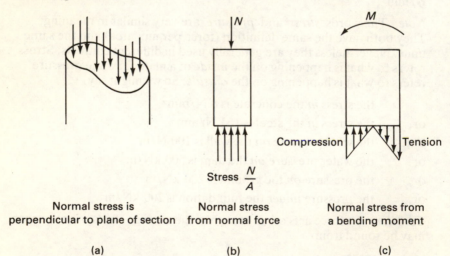

Normal stress is perpendicular to plane of section	Normal stress from normal force	Normal stress from a bending moment
(a)	(b)	(c)

Fig. 5.1 Normal stresses

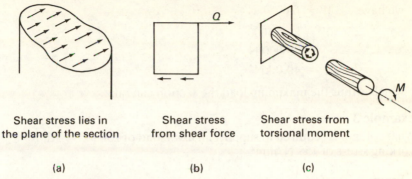

Shear stress lies in the plane of the section

Shear stress from shear force

Shear stress from torsional moment

(a)   (b)   (c)

Fig. 5.2   Shear stresses

5.2   The stress from a normal force

This is the easiest. The formula follows directly from Definition 5.1 of stress and should be learnt by heart.

Formula 5.1

Normal stress, $f = N/A$ where N is the normal force, and A is the area on which it acts.

Corollary If a load W acts on an area A then there is a normal stress on the area of, $f = W/A$

Example 1

A Universal Column 305 × 305 × 97 kg/m has a cross-sectional area of 123.3 cm². It carries 800 kN on each flange. Calculate the normal stress. If the permissible direct stress is 155 N/mm² is the column safe?

Working in units of N and mm we have,

Normal stress $f = \dfrac{W}{A}$

$$= \frac{1600 \times 10^3}{123.3 \times 10^2}$$

$$= 130 \text{ N/mm}^2$$

which is less than 155 N/mm² so the column is all right.

Example 2

Calculate the maximum axial load that a timber section 50 mm × 100 mm may take if the stress may not exceed 8 N/mm².

From $f = \dfrac{W}{A}$

we have $\qquad W = fA$

So $\qquad\qquad W = 8 \times 5000$

$\qquad\qquad\qquad = 40\ 000$ N

$\qquad\qquad\qquad = 40.0$ kN

which is thus the maximum load the section can take.

Example 3

What column is required to support an axial load of 2300 kN? Allow working stress of 155 N/mm^2.

From $\qquad f = \dfrac{W}{A}$

we have $\qquad A = \dfrac{W}{f}$

So $\qquad\quad A = \dfrac{2300 \times 10^3}{155}$

$\qquad\qquad\quad = 14\ 840$ mm^2

$\qquad\qquad\quad = 148.4$ cm^2

Hence, from the section tables we choose UC 305 $\times$ 305 $\times$ 118 kg/m which has an area of 149.8 cm^2 (see Appendix 2).

5.3 The stress from a bending moment

The importance of this section cannot be overemphasised. The formulae must be learnt and thoroughly understood. Note that the stress being considered in this section is a normal stress (perpendicular to the plane of the section considered) just like the stress of the previous section. However, because it is caused by a bending moment it is called a bending stress. The symbols in these formulae are explained in Figs. 5.3 and 5.4.

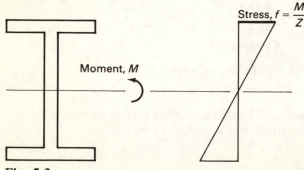

Moment, M

Stress, $f = \dfrac{M}{Z}$

Fig. 5.3

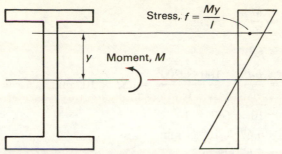

Fig. 5.4

Formula 5.2

Bending stress in outer edge of section $\quad f = \dfrac{M}{Z}$

where M is the bending moment, and Z is the section modulus.

Formula 5.3

Bending stress at distance y from the centroid $\quad f = \dfrac{My}{I}$

where I is the second moment of area.

Corollary There is no stress at the centroid of a section when a bending moment acts alone. Hence the neutral axis passes through the centroid of the section.

The proofs of these formulae are given in Appendix 3.

Example 4

The UB 406 × 178 × 74 kg/m has $Z = 1324$ cm³ (Appendix 1). What stress does it experience in its outer edges when subjected to a bending moment of 200 kN m? If the stress may be 165 N/mm² but no more, can this UB take the moment?

From $\quad f = \dfrac{M}{Z}$

we have $\quad f = \dfrac{200 \times 10^6}{1324 \times 10^3}$

$\quad = 151$ N/mm²

which is less than 165 N/mm², so it is all right.

Example 5

What bending moment can a timber section 100 × 200 mm take if the maximum allowable stress is 10 N/mm²?

From $$f = \frac{M}{Z}$$

we have $$M = fZ$$

For a rectangle $$Z = \frac{bd^2}{6} = \frac{100 \times 200^2}{6} = 667 \times 10^3 \text{ mm}^3$$

Hence
$$M = 10 \times 667 \times 10^3$$
$$= 6.67 \times 10^6 \text{ N mm}$$
$$= 6.67 \text{ kN m}$$

Example 6

What UB is required to take a bending moment of 350 kN m? Allow a maximum stress of 165 N/mm^2.

From $$f = \frac{M}{Z}$$

we have $$Z = \frac{M}{f}$$

Hence
$$Z = \frac{350 \times 10^6}{165}$$
$$= 2.12 \times 10^6 \text{ mm}^3$$
$$= 2120 \text{ cm}^3$$

So from the tables (Appendix 1) we choose UB 533 × 210 × 122 kg/m which has a section modulus of 2799 cm^3.

5.4 The stress from a shear force

This section is the most difficult of the three sections 5.2, 5.3 and 5.4 and the least used by far. However, because it is a little more complicated it is more interesting and worth the effort to understand. Figure 5.5a illustrates the formula.

Formula 5.4

Shear stress, $$q = \frac{Q}{Ib} \Sigma Ay$$

where I is the second moment of area of the whole section about the neutral axis,
b is the breadth of the section at the level for which shear stress q is required,
Q is the total shear force on the section,

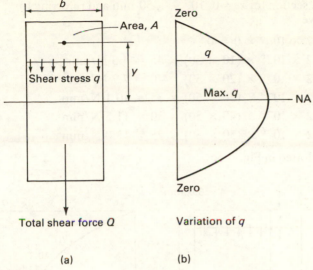

Fig. 5.5

ΣAy is the sum of the moments about the neutral axis of the areas lying above the level at which q is sought.

We have now:

Corollary 1 At the extreme edges of a section the shear stress is zero because ΣAy is zero.
Corollary 2 The shear stress will be a maximum on the neutral axis because ΣAy is a maximum there.

These two corollaries are illustrated in the variation of shear stress shown in Fig. 5.5b.

Example 7

Calculate the distribution of shear stress in a rectangular section 50×100 mm under the action of a shear force of 40 kN.

Quoting the formula,

$$q = \frac{Q}{Ib} \Sigma Ay$$

Now, $Q = 40\,000$ N, $I = bd^3/12 = 50 \times 100^3/12 = 4.17 \times 10^6$ mm^3 and $b = 50$ mm.

So $\quad \dfrac{Q}{Ib} = 192 \times 10^{-6}$

80

Working down the section for $x = 0, 10, \ldots, 50$ mm and referring to Fig. 5.6(b), we have

at $x = 0$, $\quad q = 192 \times 10^{-6} \times 0 = 0$
at $x = 10$, $\quad q = 192 \times 10^{-6} \times (10 \times 50) \times 45 = 4.3$ N/mm^2
at $x = 20$, $\quad q = 192 \times 10^{-6} \times (20 \times 50) \times 40 = 7.7$ N/mm^2
at $x = 30$, $\quad q = 192 \times 10^{-6} \times (30 \times 50) \times 35 = 10.1$ N/mm^2
at $x = 40$, $\quad q = 192 \times 10^{-6} \times (40 \times 50) \times 30 = 11.5$ N/mm^2
at $x = 50$, $\quad q = 192 \times 10^{-6} \times (50 \times 50) \times 25 = 12.0$ N/mm^2

These results are plotted in Fig. 5.6c.

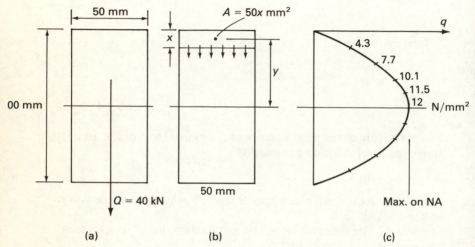

Fig. 5.6

Example 8

A built-up plate girder is shown in Fig. 5.7a. It carries a shear force of 900 kN. Calculate the maximum shear stress. How does the shear stress vary in the section?

We first obtain the second moment of area,

$$I = 3.34 \times 10^9 \text{ mm}^4$$

then we use

$$q = \frac{Q}{Ib} \Sigma Ay$$

Now at the junction of the flange and the web there is an abrupt change of b from 300 mm to 20 mm and this will result in an abrupt change in shear force by a factor of $300/20 = 15$ times.

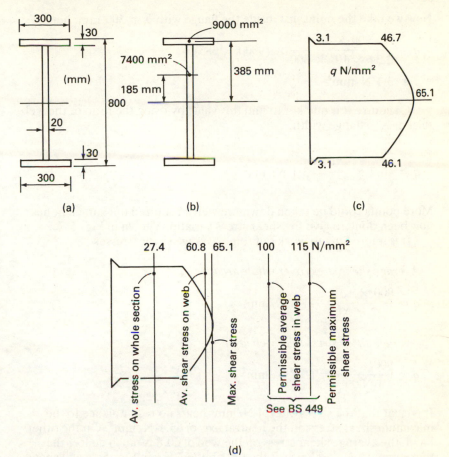

Fig. 5.7

Referring to Fig. 5.7b we obtain:
At neutral axis,

$$q = \frac{900 \times 10^3}{3.34 \times 10^9 \times 20} \times (9000 \times 385 + 7400 \times 185)$$

$$= 65.1 \text{ N/mm}^2 \quad \text{(this is the maximum shear stress)}.$$

At top of web,

$$q = \frac{900 \times 10^3}{3.34 \times 10^9 \times 20} \times (9000 \times 385)$$

$$= 46.7 \text{ N/mm}^2$$

Now we take the point just inside the flange with $b = 300$ mm; hence,

$$q = \frac{900 \times 10^3}{3.34 \times 10^9 \times 300} \times (9000 \times 385)$$

$$= 3.1 \text{ N/mm}^2$$

Of course it is quicker to find this value by using the ratio of the web thickness to flange width.

Thus,

$$q = 46.7 \times \frac{20}{300} = 3.1 \text{ N/mm}^2$$

More points could be taken down the web if required but sufficient has now been done to give the shear stress variation shown in Fig. 5.7c.

It is interesting to work out some average shear stresses.

Average shear stress over whole area

$$q = \frac{900 \times 10^3}{32\ 800} = 27.4 \text{ N/mm}^2$$

Average shear stress over web area

$$q = \frac{900 \times 10^3}{14\ 800} = 60.8 \text{ N/mm}^2$$

It is clear that the stress of 27.4 N/mm² bears no resemblance to the maximum shear stress on the neutral axis of 65.1 N/mm². On the other hand, the average shear stress on the web of 60.8 is much nearer the maximum value. A glance at the distribution of q in Fig. 5.7c will show that the web takes nearly all the shear stress and so in practice design may be based on the average stress in the web. For example, in BS 449 Tables 10 and 11 either of these conditions may be taken for mild steel: either, maximum shear stress must not exceed 115 N/mm²; or, average shear stress in web must not exceed 100 N/mm². These points are illustrated in Fig. 5.7d. Those who fixed the code have made the permissible maximum and average shear stresses further apart than the actual maximum and average shear stresses. Why was this a wise move?

5.5   Normal and bending stresses combined

The importance of this section cannot be exaggerated. As the heading suggests, it combines the effect of a normal force and a bending moment both of which produce normal stresses. The fact is that the combined effect is found by adding together the two separate effects.

Formula 5.5

The combined effects of a load W and a moment M are stresses, f, in the outer edges of the section,

where $$f = \frac{W}{A} \pm \frac{M}{Z}$$

The sign to be taken is determined by the direction of the bending moment M (Figs. 5.8a, b, c).

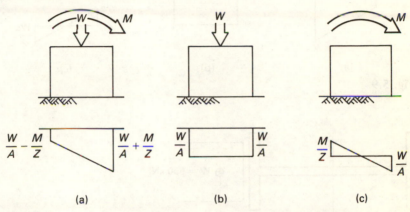

(a)   (b)   (c)

Fig. 5.8

Corollary A load W applied eccentrically a distance e from the axis of bending will produce stresses, f, in the outer edges of the section,

where $$f = \frac{W}{A} \pm \frac{We}{Z}$$

This corollary is very important but quite simple. It will arise in practice in many different situations: a column with a load on one side; a foundation with a column off-centre; a prestressed concrete beam with an eccentric prestressing force – all need this corollary for their solution.

If a load is eccentric, at distance e from the axis of bending (which passes through the centroid) then bending is introduced as well as axial compression. Figure 5.9 shows how the two effects can be analysed separately and then combined as in the corollary.

Example 9 (Fig. 5.10)

The UC 356 × 368 × 129 kg/m supports a load of 650 kN off one flange. Determine the stresses in both flanges.

Taking the section properties from the tables (Appendix 2),

$$A = 164.9 \times 10^2 \text{ mm}^2, \qquad Z = 2264 \times 10^3 \text{ mm}^3$$

84

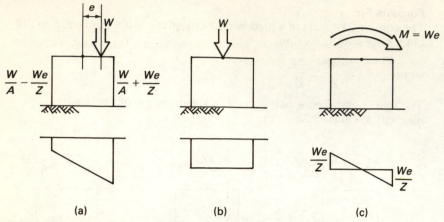

$$\frac{W}{A} - \frac{We}{Z} \qquad\qquad \frac{W}{A} + \frac{We}{Z}$$

$$M = We$$

$$\frac{We}{Z} \qquad\qquad \frac{We}{Z}$$

(a)   (b)   (c)

Fig. 5.9

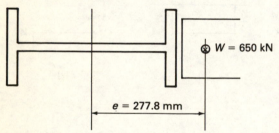

$W = 650$ kN

$e = 277.8$ mm

Fig. 5.10

Hence,

axial stress, $\qquad\qquad f = \dfrac{W}{A} = \dfrac{650 \times 10^3}{164.9 \times 10^2} = 39.4 \text{ N/mm}^2$

and

bending stress, $\qquad f = \dfrac{We}{Z} = \dfrac{650 \times 10^3 \times 277.8}{2264 \times 10^3} = 79.8 \text{ N/mm}^2$

So,

Combined stress, $\qquad f = \dfrac{W}{A} \pm \dfrac{We}{Z} = 39.4 \pm 79.8$

$$= 119.2, \quad \text{and} \quad -40.4 \text{ N/mm}^2$$

that is, 119.2 N/mm^2 compression in the flange nearer to the load and
40.4 N/mm^2 tension in the flange further from the load.

Example 10 (Fig. 5.11)

A prestressed concrete beam 300 mm × 900 mm has a prestressing force of $H = 1350$ kN acting 150 mm below the neutral axis. Determine the stress distribution in the section.

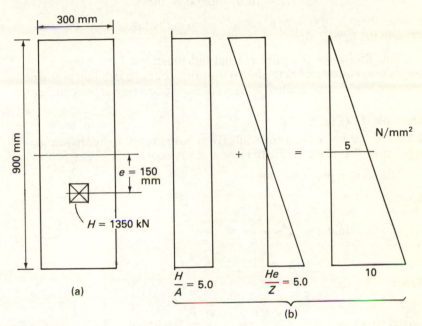

Fig. 5.11

The section properties are easily found.

Area, $A = 300 \times 900 = 270 \times 10^3$ mm^2

section modulus, $Z = \dfrac{bd^2}{6} = \dfrac{300 \times 900^2}{6}$

$= 40.5 \times 10^6$ mm^3

Hence,

axial stress, $\qquad f = \dfrac{H}{A} = \dfrac{1350 \times 10^3}{270 \times 10^3}$

$= 5.0$ N/mm^2

and

bending stress, $\qquad f = \dfrac{He}{Z} = \dfrac{1350 \times 10^3 \times 150}{40.5 \times 10^6}$

$= 5.0$ N/mm^2

So,

Combined stress, $\quad f = \dfrac{H}{A} \pm \dfrac{He}{Z} = 5.0 \pm 5.0$

$$= 10.0, \text{ and } 0.0 \text{ N/mm}^2$$

That is, there is 10.0 N/mm^2 compression in the lower edge, and no stress in the upper edge.

This illustrates the principle behind prestressed concrete: a longitudinal force arches the beam upwards so that a bending moment from the load can be applied with no tension occurring in the concrete.

Example 11 (Fig. 5.12)

A load of 100 kN acts eccentrically from both axes of a square foundation 2 m $\times$ 2 m. Determine the stress at each corner.

The section properties are:

Area, $A = 4 \text{ m}^2$

section modulus, $\quad Z = \dfrac{bd^2}{6} = \dfrac{2 \times 2^2}{6}$

$$= 1.333 \text{ m}^3$$

Hence,

axial stress, $\qquad f = \dfrac{W}{A} = \dfrac{1000}{4} = 250 \text{ kN/m}^2$

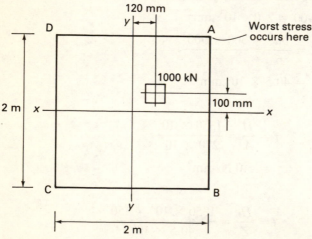

Fig. 5.12

Also,

$$\text{bending stress } (X - X), \quad f = \frac{We}{Z} = \frac{1000 \times 0.1}{1.333}$$
$$= 75 \text{ kN/m}^2$$

which is compression along DA and tension along CB.
Similarly,

$$\text{bending stress } (Y - Y), \quad f = \frac{We}{Z} = \frac{1000 \times 0.120}{1.333}$$
$$= 90 \text{ kN/m}^2$$

which is compression along AB and tension along DC.

Taking compression as positive, we obtain the following combined stresses:

Stress at
$$A = 250 + 75 + 90 = 415 \text{ kN/m}^2$$
$$B = 250 - 75 + 90 = 265 \text{ kN/m}^2$$
$$C = 250 - 75 - 90 = 85 \text{ kN/m}^2$$
$$D = 250 + 75 - 90 = 235 \text{ kN/m}^2$$

5.6 Normal and shear stresses combined

Look at Fig. 5.13a. At any section along the beam there is a shear force and a bending moment. The shear force will produce shear stresses and the bending moment will produce normal stresses. If we take a typical point in the beam – say, between mid span and the support and between

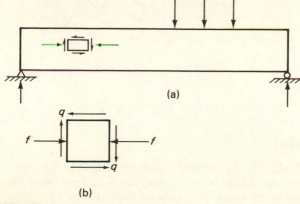

(a)

(b)

Fig. 5.13

88

the neutral axis and the top edge – then shear stresses and normal stresses will be found together (Fig. 5.13b). Note the way the shear stresses act around the material. It may not seem obvious that there are shear stresses on the horizontal planes – but bend a pack of cards and you will see that the cards slide over each other which shows the existence of shear stresses in the plane of the cards.

We need now to turn to the analysis of this type of stress combination. For a first reading the next two sections may be omitted.

5.7 Principal stresses

Let us consider the general state of stress as shown in Fig. 5.14a. We adopt the sign convention that tension is positive for the normal stresses, and if the shear stress runs clockwise round the material then that is taken as positive also. The question arises, what is the state of stress at the same point but on different planes (Fig. 5.14b)? Furthermore, what are the maximum and minimum normal stresses and on which planes do they act (Fig. 5.14c)? As it turns out these maximum and minimum normal stresses occur on planes at right angles to each other and on which the shear stress is zero.

To work all this out requires hard work resolving some forces and solving some equations . . . but a very neat geometrical construction avoids all this labour. It is called Mohr's circle after the man who thought it out, and we turn to it now.

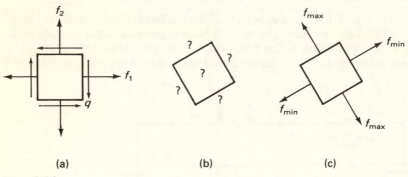

(a) (b) (c)

Fig. 5.14

5.8 Mohr's circle

Consider again Fig. 5.14a. On the vertical planes there are the stresses f_1 and q and on the horizontal planes the stresses are f_2 and $-q$ (note the minus sign because the shear stress runs anticlockwise round the material). Now suppose we represent these pairs of stresses, $V(f_1, q)$

and $H(f_2, -q)$, by two points referred to a pair of axes (Fig. 5.15a). Joining the points to cut the axis in O gives the circle centre O passing through these two points (Fig. 5.15b). Now the interesting fact about this circle is that any point on it will give the stresses in another direction. The vital question is – on what planes do these other stresses act? To answer this, draw a vertical line through the point corresponding to the stresses on the vertical plane, V, and a horizontal line through H to meet the circle at the pole, P (Fig. 5.15c). Now lines through P will give the planes on which are acting the stresses at the other end of the line. Consider for example the principal stresses p_1 and p_2 – these are the maximum and minimum normal stresses (Fig. 5.15d). Simply join up the pole to the points where the circle cuts the f axis, A and B. The maximum principal stress, p_1, then acts on a plane parallel to PA and p_2 acts on the plane parallel to PB. Note that the properties of the circle ensure that PA and PB are at right angles to each other.

On which planes are the shear stresses a maximum? Are the normal stresses zero on these planes?

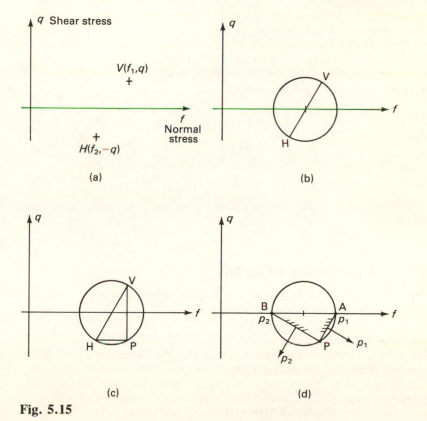

Fig. 5.15

90

Example 12 (Fig. 5.16)

At a particular point in a structure the shear and normal stresses are as shown in Fig. 5.16a. Determine the principal stresses.

Draw Mohr's circle and the pole and join the pole to the points A, B where the circle cuts the normal stress axis. By measurement the principal stresses are,

$p_1 = 20.4$ N/mm² and $p_2 = 1.6$ N/mm² acting on planes at 61° to the horizontal and vertical respectively (Fig. 5.16b, c).

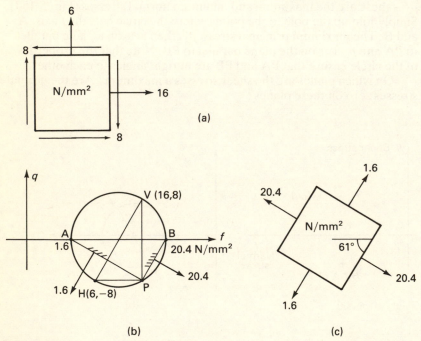

(a)

(b)

(c)

Fig. 5.16

5.9 The definition of strain

A quantity associated with stress – but quite different from it – is strain. Its definition must be learnt by heart.

Definition 5.4

STRAIN is the increase in length per unit length. Since it is a ratio it has no units. It may be calculated from

$$\text{Strain,} \quad \varepsilon = \frac{\text{increase in length } (e)}{\text{original length } (L)}$$

This type of strain comes from the application of normal stresses and forces. There is such a thing as shear strain (and other types of strain too), but we are not concerned with them here.

The connection between stress (f) and strain (ε) is also important. For most materials the strain is doubled if the stress is doubled – at least if the stresses are not too high.

This means that stress/strain is a constant. This constant has the name Young's modulus, E, and its units are the same as those of stress, usually N/mm^2 or kN/mm^2. Expressing this as a definition we have:

Definition 5.5
The YOUNG'S MODULUS, E, of a material is its stress divided by its strain.

So,   Young's modulus, $E = \dfrac{\text{stress}}{\text{strain}}$

All this presupposes that the stresses do not go too high. If they do then the extensions and hence the strains increase much more rapidly than the stresses and so the stress/strain ratio is no longer constant.

5.10   Normal stress and strain in steel

The previous section will now be illustrated for the case of steel. Suppose a vertical steel wire 4 m long and 1 mm diameter is stretched by attaching a mass to its free end. Recording the extension of the wire makes it possible to calculate the stress and strain and hence the Young's modulus. Here is a typical result:

mass = 8.0 kg,   extension = 2.0 mm

Hence,

stress, $f = \dfrac{\text{weight}}{\text{area}} = \dfrac{8 \times 9.81}{0.785}$

$= 100 \ N/mm^2$

and

strain, $\varepsilon = \dfrac{\text{extension}}{\text{original length}} = \dfrac{2.0}{4000}$

$= 500 \times 10^{-6}$

So,

Young's modulus, $E = \dfrac{f}{\varepsilon} = \dfrac{100}{500 \times 10^{-6}}$

$= 200\ 000 \ N/mm^2$

If the experiment were repeated for different weights then the stress–strain graph would be a straight line (Fig. 5.17a) whose slope was the Young's Modulus, E. After further loading, the wire yields at a constant stress called the yield stress, $f_y = 250$ N/mm² for mild steel, but amazingly hardens up again so that the wire can again take more load (Fig. 5.17b). Contracting the scale on the strain axis by one-tenth and further increasing the load shows that there is still a long way to go before the wire snaps (Fig. 5.17c).

Example 13

A member of a steel truss is 4 m long and has a force in it of 100 kN. Calculate the extension of the member if its cross-sectional area is 1000 mm².

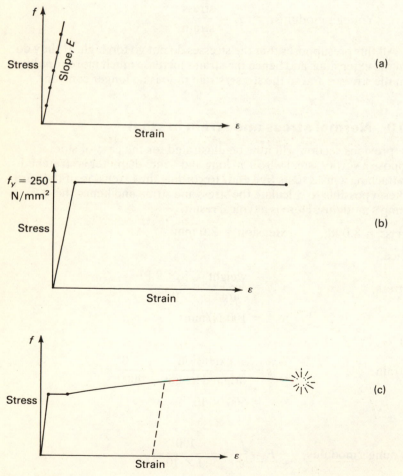

Fig. 5.17

We have

stress, $\quad f = \dfrac{\text{load}}{\text{area}} = \dfrac{100\ 000}{1000} = 100\ \text{N/mm}^2$

From

$$\frac{f}{\varepsilon} = E$$

we obtain

strain, $\quad \varepsilon = \dfrac{f}{E} = \dfrac{100}{200\ 000} = 0.5 \times 10^{-3}$

From,

$$\varepsilon = \frac{e}{L}$$

we now find

extension, $\quad e = \varepsilon L$
$\qquad\qquad\quad = 0.5 \times 10^{-3} \times 4000$
$\qquad\qquad\quad = 2\ \text{mm}$

5.11 Normal stress and strain in concrete

Concrete is not nearly so well behaved as steel. Its properties are far more variable and also it does not have a straight-line stress – strain relationship. Fig. 5.18 gives a typical curve. The initial slope will give the Young's modulus for concrete, which is about $E = 25\ 000$ to $30\ 000\ \text{N/mm}^2$. This range itself shows how much concrete can vary in

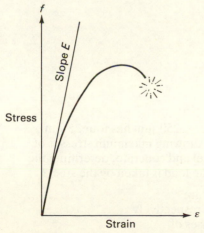

Fig. 5.18

its properties. It is further complicated by the fact that concrete shrinks on setting, creeps under load and increases in strength with age. This does not mean it is not a useful material – it is – it just means that it is difficult to analyse its behaviour accurately. Of course, all the formulae to do with stress or strain in this chapter can be assumed to be all right as a first approximation to the behaviour of concrete.

5.12 Normal stresses in members composed of two materials

Often a part of a structure may be made of two materials acting together. An example is a reinforced concrete column. Both the concrete and the steel reinforcement share the load. Here is the key to analysing such a situation:

When two materials act together
then
the strain in each is the same

So,

$$\frac{f_s}{E_s} = \frac{f_c}{E_c} \qquad \text{(s for steel, c for concrete)}$$

Rearranging,

$$f_s = \frac{E_s}{E_c} \times f_c$$

In round figures we may take

$$E_s/E_c = 200\,000/25\,000 = 8$$

Hence,

$$f_s = 8f_c$$

Or, finally,

Steel stress $\doteq 8 \times$ concrete stress

Example 14

A reinforced concrete column 250 mm × 250 mm has four 25 mm diameter mild-steel reinforcing bars. Allowing maximum stresses of 140 N/mm^2 and 10 N/mm^2 for the steel and concrete, determine the load it can take. What percentage of the load is taken by the steel?

Clearly the steel and concrete areas are
$A_s = 1963$ mm^2, $A_c = 60\,537$ mm^2 respectively
Taking $E_s/E_c = 8$ we have the stresses of
$f_s = 80$ N/mm^2, $f_c = 10$ N/mm^2

and hence

    steel force $= f_s A_s = 157\,000$ N $= 157$ kN
and concrete force $= f_c A_c = 605\,000$ N $= 605$ kN
Hence, the column may take a total of 762 kN.
    Clearly the percentage of this total taken by the steel is

$$\frac{157}{762} \times 100 \doteq 20 \text{ per cent.}$$

5.13 Summary of formulae

Normal stress $f = \dfrac{W}{A}$

Bending stress $f = \dfrac{M}{Z}$ $f = \dfrac{My}{I}$

Shear stress $q = \dfrac{Q}{Ib} \Sigma Ay$

Normal stress and bending stress $f = \dfrac{W}{A} \pm \dfrac{M}{Z}$ $f = \dfrac{W}{A} \pm \dfrac{We}{Z}$

Strain $\varepsilon = \dfrac{e}{L}$

Stress, strain, Young's modulus $\dfrac{f}{\varepsilon} = E$

Stresses in two materials $f_2 = \dfrac{E_2}{E_1} \times f_1$

5.14 Exercises

1 Explain the difference between force and stress, giving one clear
   example of each.
2 Define normal stress and quote the formula for it.
3 A Universal Column has a cross-sectional area of 123.3 cm^2 and
   carries an axial load of 1200 kN. Calculate the stress. If the
   permissible stress is 120 N/mm^2, can the column support the load?
4 What Universal Column is required to support an axial load of
   800 kN if the stress must not exceed 100 N/mm^2?
5 Determine the axial load that the smallest UC can support without
   the stress exceeding 155 N/mm^2.
6 A square foundation measures 1.6 m $\times$ 1.6 m and supports a

centrally placed load of 500 kN. Determine the pressure (the same as stress) under the foundation. Is the foundation safe if the ground can take 200 kN/m^2?

7 Calculate the size of a square foundation which can safely carry 1000 kN on to ground whose permissible bearing pressure must not exceed 130 kN/m^2.

8 Quote the two formulae which will give the stress in a section due to a bending moment. Explain clearly with a sketch where the bending stresses are and which stresses the formulae give.

9 A timber joist 100 mm × 200 mm has a permissible bending stress of 8N/mm^2. Calculate the maximum bending moment that it can safely withstand.

10 A Universal Beam has I = 45 717 cm^4 and an overall depth D = 467.4 mm. Determine the stress 100 mm from the neutral axis due to a bending moment of 325 kN m.
(*Hint:* work in N and mm units; 1 cm^4 = 10^4 mm^4)

11 What UB is required to resist a bending moment (BM) of 250 kNm if the stress must not exceed 165 N/mm^2?

12 What BM can the smallest UB resist? Allow a permissible stress of 165 N/mm^2.

13 What BM can the largest UB resist? Again, allow 165 N/mm^2.

14 A section is formed from three plates: the web plate is 500 mm × 30 mm and the two flanges 200 mm × 30 mm each. The second moment of area is I = 1156 × 10^6 mm^4. The whole section carries a shear force Q = 1500 kN. Determine the variation in shear stress in the section. Compare the average and the maximum stresses with those allowed in BS 449 (Fig. 5.7d).

15 A T-section with 100 mm × 10 mm flange and 200 mm × 10 mm web carries a shear force of 120 kN. Plot the shear stress distribution.

16 A square foundation 2 m × 2 m carries a central load of 500 kN and a bending moment of 60 kN m about an axis parallel to one side. Sketch the pressure distribution under the base. To what value may the moment increase before there is uplift?

17 A timber column 300 mm × 300 mm carries a vertical load of 100 kN eccentric 100 mm from one axis. Does any tension occur? Sketch the stress distribution giving the values of the worst stresses.

18 The smallest Universal Column supports a load of 80 kN off one flange with eccentricity e = 176.2 mm from the X–X axis. Determine the stress distribution and state the maximum compression and tension that will occur.

19 At a certain point in a steel beam there is a longitudinal normal stress of 100 N/mm^2 tension and a shear stress of 50 N/mm^2. Draw Mohr's circle and determine the principal stresses, showing clearly the planes on which they act.

20 In a prestressed concrete beam there is at a certain point a longitudinal compression of 10 N/mm^2 and a shear stress of 2 N/mm^2. Draw a stress block with these stresses marked on it. Using Mohr's

circle determine the maximum principal tension that occurs and show clearly the direction in which it is acting.

21 At a certain point in a concrete dam there is a vertical normal stress of 8.5 N/mm² compression, a horizontal normal stress of 0.5 N/mm² compression and a shear stress of 3 N/mm². Draw Mohr's circle for this state of stress. Does tension occur in the concrete in any direction?

22 Define strain. What are the units of strain?

23 Distinguish between stress and strain. Quote the formula which connects them, explaining the symbols you use.

24 Define Young's modulus. What are its units? What is the Young's modulus for (a) steel and (b) concrete?

25 A prestressed concrete beam of 20 m span is stressed by tensioning one hundred 5 mm diameter high tensile strength steel wires to give a total force of 1400 kN. Determine the stress in the wires. Using a Young's modulus of 200 kN/mm² deduce the strain and the extension of wires.

26 The cable of a suspension bridge has a stress in it of 650 N/mm² and its length is 1000 m. Determine the strain in the cable and its extension. Take a Young's modulus of 200 kN/mm².

27 A circular concrete column of 300 mm diameter is reinforced with six 25 mm diameter reinforcing bars. The steel stress may not exceed 140 N/mm² and the concrete stress is 12 N/mm². What is the stress, strain and load in (a) the concrete and (b) the steel? Hence deduce the permissible load of the column. You may assume the Young's moduli for steel and concrete are $E_s = 200\ 000$ N/mm² and $E_c = 25\ 000$ N/mm² respectively.

Part II

The basic structures

Chapter 6

Beams and cantilevers

6.1 The support reactions

It is nearly always essential to begin the analysis of a beam by finding the reactions. They can be found very simply by using the two basic principles of equilibrium – resolving forces in any direction and taking moments about any point.

How to find the reactions
1 Find the *left* reaction by taking moments about the right-hand support.
2 Find the *right* reaction by resolving vertically.

Example 1 (Fig 6.1)
Calculate the reactions R_1 and R_2.

Taking moments about B,

clockwise moments = anticlockwise moments

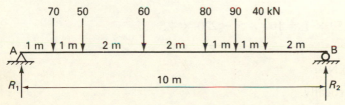

Fig. 6.1

Therefore,

$$R_1 \times 10 = 70 \times 9 + 50 \times 8 + 60 \times 6 + 80 \times 4 + 90 \times 3 + 40 \times 2$$

Hence,

$$R_1 = 206 \text{ kN}$$

Resolving vertically,

forces up = forces down

So,

$$R_1 + R_2 = 70 + 50 + 60 + 80 + 90 + 40$$

Hence,

$$R_2 = 184 \text{ kN}$$

Example 2 (Fig. 6.2)

Here we have a uniformly distributed load (UDL) and overhanging ends but the principle is the same as before. It is best to think of the UDL acting with its total value ($20 \times 11 = 220$ kN) at its centre of gravity, G, which is at the centre of its whole length – that is 5.5 m from either end but 4.5 m from the right support. And don't forget the 50 kN is acting clockwise about B.

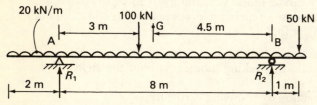

Fig. 6.2

Taking moments about B,

clockwise moments = anticlockwise moments

So,

$$R_1 \times 8 + 50 \times 1 = 100 \times 5 + 220 \times 4.5$$

Hence,

$$R_1 = 180 \text{ kN}$$

Resolving vertically,

forces up = forces down

So,

$R_1 + R_2 = 100 + 220$

Hence,

$R_2 = 140$ kN

The cantilever is a little different because it only has one support. The reaction at this support consists of a vertical force V and a moment M and by resolving vertically and taking moments about the support we can easily find them.

Example 3 (Fig. 6.3)

Calculate the reactions for this cantilever.

Resolving vertically,

forces up = forces down

So

$V = 12 + (15 \times 4)$

that is,

$V = 72$ kN

Taking moments about the support,

anticlockwise moments = clockwise moments

Hence,

$M = 12 \times 3 + (15 \times 4) \times 2$

So

$M = 156$ kN m   (note the units − not kN/m!)

The examples chosen so far have only had vertical loads. If an inclined load acts it will mean that a horizontal force will appear at a support. In the case of the beam one support must be capable of taking a horizontal force and the other must be on a roller where there will be no horizontal force.

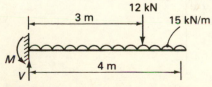

Fig. 6.3

Example 4 (Fig. 6.4)

Determine the reactions for the cantilever over the end of which runs a cable carrying a 10 kN tensile force.

Working with the components of the forces,

Resolving horizontally,

forces to the right = forces to the left

Hence,

$$H = 10 \cos 40 + 10 \cos 65$$

So

$$H = 11.89 \text{ kN}$$

Resolving vertically,

forces up = forces down

Therefore,

$$V + 10 \cos 50 = 10 \cos 25$$

So

$$V = 2.64 \text{ kN}$$

Finally, taking moments about the support,

anticlockwise moments = clockwise moments

So

$$M + 10 \cos 50 \times 7 = 10 \cos 25 \times 7$$

Therefore,

$$M = 18.4 \text{ kN m}$$

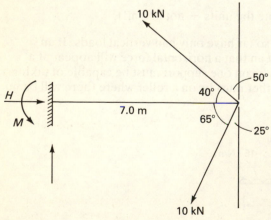

Fig. 6.4

6.2   The shear force diagram

This is an important diagram to be able to sketch quickly and correctly.
It follows directly from the reactions. Look at Fig. 6.5a. The shear
force, F, at a certain point acts as shown. Remember that to find an
internal force the structure must be cut into two parts and attention
fixed on one half – the left half in this case.

Resolving vertically,
shear force, F = reactions − loads

   This means that the shear force is initially (at the left of the beam)
equal to the left-hand reaction and then it reduces by the value of the
loads in moving across the beam (Fig. 6.5b). Or we may say the shear
force (SF) diagram moves up at the reactions and down at the loads – in
other words, the SF diagram follows the forces on the beam.

How to draw the shear force diagram
Follow the forces on the beam from the left-hand end.

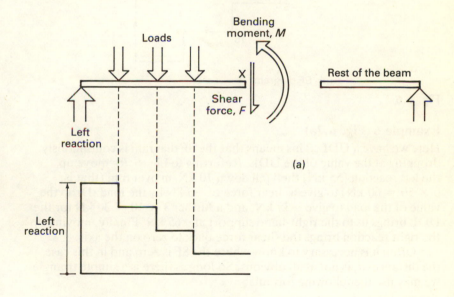

(b)   Shear force diagram

Fig. 6.5

Example 5 (Fig. 6.6a)

Clearly the reactions are each 25 kN. Look now at Fig. 6.6b. Starting
from the left we move up the left-hand reaction, along to the first load,
down the value of the load (10 kN), along to the next load, and so on.

Note that at the right-hand end we move up the right-hand reaction to land back at zero again. This must always happen – if it does not you have made a mistake!

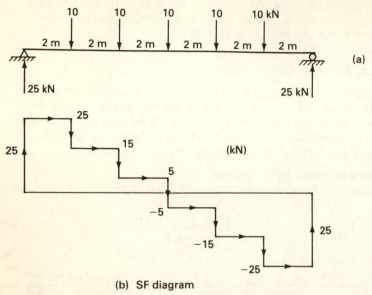

(a)

(b) SF diagram

Fig. 6.6

Example 6 (Fig. 6.7a)

Here we have a UDL. This means that the SF diagram is continuously dropping at the value of the UDL. Referring to Fig. 6.7b: move up the left reaction (55 kN) then fall down 10 kN/m over 6 m (that is, $6 \times 10 = 60$ kN) to give a shear force of −5 kN by the load. Drop the value of the load to give −35 kN, and a further $3 \times 10 = 30$ kN for the UDL brings us to the right-hand support at −65 kN. Finally, moving up the right reaction brings the shear force back to zero on the axis.

Often it is necessary to know where the SF is zero and in this case the distance, d, is not at all obvious. So long as there is a simple triangle we may use the following formula:

Formula 6.1
Distance to zero shear force

$$d = \frac{\text{shear force}}{\text{UDL}}$$

In our case (Fig. 6.7b),

$$d = \frac{55}{10} = 5.5 \text{ m from the left support.}$$

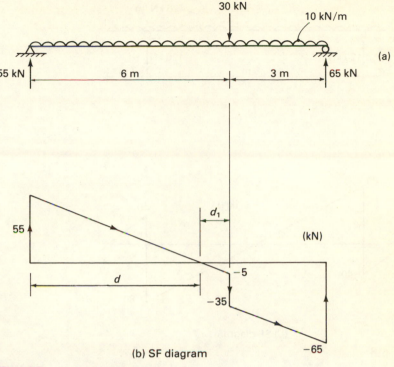

Fig. 6.7

But we could have taken the other triangle,

$$d_1 = \frac{5}{10} = 0.5 \text{ m to the left of the point load}$$

which is of course the same point.

Example 7 (Fig. 6.8a)

Here there is an overhanging end – but the principles stay the same: find the reactions then follow the forces. The SF diagram is shown in Fig. 6.8b. Do you see how to obtain these figures?

6.3 The bending moment diagram

Generally speaking, the beam that is required to support a given set of loads is chosen so that it can withstand the maximum bending moment. It is the bending moments which are the most crucial. If a beam can take the bending moments then it is quite likely that it can take the shear forces. The variation in the bending moments across the beam – as shown in the bending moment diagram – is therefore of first importance.

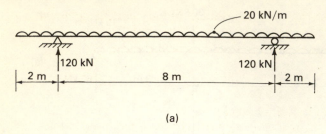

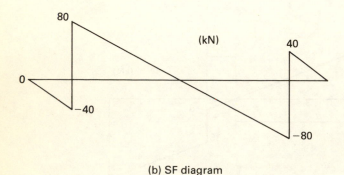

(a)

(b) SF diagram

Fig. 6.8

Consider again Fig. 6.5a. When the beam is cut the internal bending moment, M, is revealed (as well as the shear force; but we have dealt with that). Let us take moments about X for the left half of the beam:

anticlockwise moments = clockwise moments

So,

M + moment of the loads = moment of the reaction

Solving for M,

bending moment, M = moment of reaction *minus* moment of loads

The question arises – at which points should the BM (bending moment) be calculated? Of course one essential point is where it is a maximum (this, it will turn out, is where the SF is zero). Other points to choose would be under the point loads; or again at equal distances (say 1 m intervals) across the span. Let us summarise like this:

How to draw the bending moment diagram
M = *reaction moment* − *load moment*
for forces to the left of the point where you want the bending moment.

There is an interesting relationship between the SF diagram and the

BM diagram which can often be useful in finding bending moments. Figure 6.9 shows part of a beam and its SF and BM diagram.

We have,

M = reaction moment − load moment

= area $(abcd)$ − area $(a_1\,b_1\,c_1\,d_1)$

= area under the SF diagram

A little more thought will extend this to the general statement:

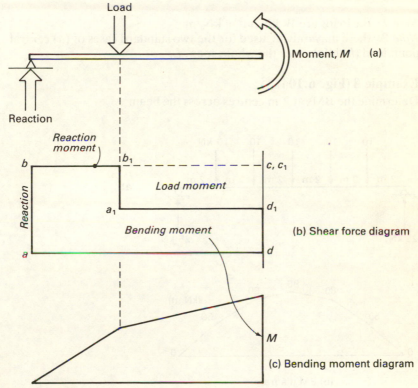

Fig. 6.9

Link between SF and BM
Increase in BM = Area under SF diagram
between any two points of a beam.

From this it follows that the BM will begin to go down when the SF goes negative. Hence we may deduce the very important conclusion:

Maximum BM *occurs at the point of* **zero SF**

Two important standard cases for the bending moment at midspan need to be well learnt.

110

Formula 6.2

Central point load $M = \dfrac{WL}{4}$

Formula 6.3

Uniformly distributed load (UDL) $M = \dfrac{wL^2}{8}$

Note 1: the loads are W kN but w kN/m.
Note 2: these may only be used for the two standard cases of (*a*) central point load (*b*) UDL over the whole span, L.

Example 8 (Fig. 6.10a)
Determine the BMs at 2 m centres across the beam.

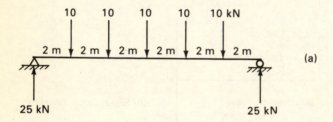

(a)

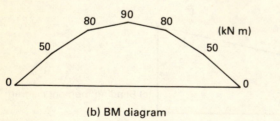

(b) BM diagram

Fig. 6.10

At left support $M = 0$
2 m from LHS, $M = 25 \times 2 = 50$ kN m   (*not* kN/m!)
4 m from LHS, $M = 25 \times 4 - 10 \times 2 = 80$ kN m
6 m from LHS, $M = 25 \times 6 - 10 \times 4 - 10 \times 2 = 90$ kN m

   Hence we may draw the BM diagram (Fig. 6.10b). It may easily be shown that the areas under the SF diagram (Fig. 6.6b) will yield the same values. *Can you do it?*

Example 9 (Fig. 6.11a)

Determine the BMs at 1 m centres across the span.

Using
M = reaction moment − load moment

We obtain:

At $x = 0$, $M = 0$

$x = 1$, $M = 100 \times 1 - (20 \times 1) \times 0.5 = 90$ kN m

$x = 2$, $M = 100 \times 2 - (20 \times 2) \times 1.0 = 160$ kN m

$x = 3$, $M = 100 \times 3 - (20 \times 3) \times 1.5 = 210$ kN m

$x = 4$, $M = 100 \times 4 - (20 \times 4) \times 2.0 = 240$ kN m

$x = 5$, $M = 100 \times 5 - (20 \times 5) \times 2.5 = 250$ kN m

Hence the BM diagram (Fig. 6.11b).

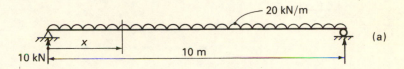

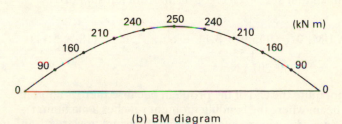

(b) BM diagram

Fig. 6.11

Example 10 (Fig. 6.12a)

Determine the maximum bending moment.

First find the left-hand reaction,
Taking moments about B,

$R_1 \times 8 = 60 \times 6 + (18 \times 9) \times 3.5$

So

$R_1 = 115.9$ kN

Second, draw the SF diagram until it is zero (Fig. 6.12b) where,

$$d = \frac{SF}{UDL} = \frac{19.9}{18} = 1.11 \text{ m}$$

112

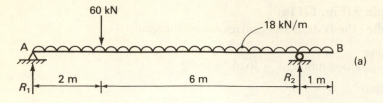

60 kN

18 kN/m

A   B

(a)

R_1 2 m   6 m R_2 1 m

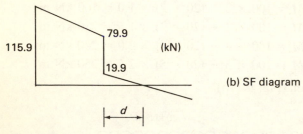

115.9   79.9   (kN)

19.9

(b) SF diagram

d

Fig. 6.12

So BM is a maximum at $2 + 1.11 = 3.11$ m from left support. Hence,

maximum BM, $M = 115.9 \times 3.11 - 60 \times 1.11 - (18 \times 3.11)$
$$\times 1.55$$
$$= 207 \text{ kNm}$$

This is not quite the end of the problem because there is another place in the beam where the bending moment reaches a maximum value – this time a hogging moment arching the beam upwards – and it occurs over the right support where the SF will again pass through zero.

Taking moments about B for the left half of the beam,
$M =$ reaction moment − load moment
$$= 115.9 \times 8 - 60 \times 6 - (18 \times 8) \times 4$$
$$= -8.8 \text{ kN m}$$

The negative sign means a hogging moment.

This moment could of course have been found other ways. For example, the moment of the UDL on the cantilever about B is $(18 \times 1) \times 0.5 = 9.0$ kN m (hogging). This is the exact value, whereas the 8.8 is a little out because the reaction has been rounded up from its exact value of 115.875 kN. Again we could have taken the net area of the SF between the left support and the right to give the moment at B again as -9.0 kN m.

Example 11 (Fig. 6.13a)

Draw the SF and BM diagrams for this cantilever.

For the SF diagram we simply follow the forces from the left to give Fig. 6.13b. And taking moments from the left gives the BM diagram (Fig. 6.13c). Can you obtain these figures?

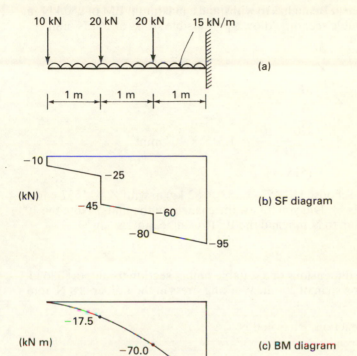

(a)

(b) SF diagram

(c) BM diagram

Fig. 6.13

6.4   The choice of section

Now comes the next part – choosing a section which will support the given loads. So long as we are dealing with a beam of one material – anything from timber to steel (but *not* reinforced concrete) – then there is a permissible bending stress which must not be exceeded.

How to choose the section size

1 Solve $\quad f = \dfrac{M}{Z} \quad$ for Z,

where f is made equal to the permissible stress, and M is made equal to the maximum bending moment.

2 Choose a section with at least this value of Z.

Example 12

A steel Universal Beam has to withstand a maximum BM of 250 kN m. Choose a suitable section. Allow a permissible stress of 165 N/mm^2.

From
$$f = \frac{M}{Z}$$

we have
$$Z = \frac{M}{f}$$

$$= \frac{250 \times 10^6}{165} = 1.515 \times 10^6 \text{ mm}^3$$

$$= 1515 \text{ cm}^3$$

Hence we choose UB 457 $\times$ 152 $\times$ 82 kg/m with $Z = 1557$ cm^3 (see Appendix 1). Did you follow the changes in the units? Note the 10^6 to change kN m to N mm and the 10^3 to change mm^3 to cm^3.

Example 13

Calculate the dimensions of a suitable timber section to support a load of 800 N over a span of 2.4 m. Working stress in the timber is 8 N/mm^2.

Clearly,
left-hand reaction, $R_1 = 400$ N

Hence,

maximum BM, $M = 400 \times 1.2 = 480$ N m (note units).

From

$$f = \frac{M}{Z}$$

we obtain

$$Z = \frac{480 \times 10^3}{8}$$

$$= 60\,000 \text{ mm}^3$$

But

$$Z = bd^2/6$$

Trying $b = 50$ mm, $d = 100$ mm gives $Z = 83\,000$ mm^3 which is more than adequate.

Again: remember to take great care with the units. Always use N and mm when substituting into $f = M/Z$. How many N mm is 626 kN m? How many mm^3 is 1073 cm^3? Do you really understand? Of course, for Universal Beams always use Z_{xx} and not Z_{yy}.

Example 14

The smallest UB ($Z = 231.9$ cm^3) spans 3.2 m. Investigate whether it is safe for it to carry a load of 50 kN at midspan.

Clearly,

maximum BM, $M = 25 \times 1.6 = 40$ kN m

Hence,

$$\text{maximum stress}, f = \frac{M}{Z} = \frac{40 \times 10^6}{231.9 \times 10^3} = 172 \text{ N/mm}^2$$

This stress is greater than the 165 N/mm^2 allowed in steel, so the beam is not suitable for the given loading.

6.5 Checking the shear stress

It is not normally necessary to calculate the actual variation in shear stress when dealing with UB sections or an I section built up from rectangular plates. BS 449 allows the average shear stress in the web to be calculated and limits its value to 100 N/mm^2 for mild steel.

Formula 6.4

Average shear stress in web $\quad q_{av} = \dfrac{Q}{Dt}$

must be less than 100 N/mm^2 for mild steel

where Q is the shear force
D is the depth of the UB
t is the thickness of the web.

Note: for a built up section D must be taken as the depth of the web and not the depth of the section.

Of course, timber sections are generally rectangular ($b \times d$) and the maximum shear stress is then known to be $1.5(Q/bd)$. CP 112 Table 4 gives the permissible shear stress for the various softwoods and hardwoods. Roughly speaking, the softwood permissible shear stresses lie between 0.5 and 1.5 N/mm^2.

Formula 6.5

Maximum shear stress in rectangular sections $\quad q = 1.5\left(\dfrac{Q}{bd}\right)$

6.6 The deflection of a beam

It is often necessary to check that the deflection of a beam is not excessive. Generally speaking the bending moment is the critical factor; if the beam can take the bending moment then more often than not the shear stresses and the deflection will not be too great. But it is not always so and therefore the deflection (like the shear) needs to be calculated.

Deflection calculations tend to be complicated, so a clever way has been devised to simplify the work. Let us begin with two formulae which are for standard cases only. They should be memorised.

Formula 6.6

Central point load, central deflection $y = \dfrac{WL^3}{48EI}$

Formula 6.7

UDL central deflection $y = \dfrac{5wL^4}{384EI}$

where, L is the span

E is the Young's modulus $\begin{cases} \text{Steel } 200 \text{ kN/mm}^2 \\ \text{Concrete, } 25 \text{ kN/mm}^2 \\ \text{Wood, } 10 \text{ kN/mm}^2 \end{cases}$

I is the second moment of area

and, W is the point load (kN)

but w is the UDL (kN/m)

and the most difficult part is to use the correct units. (Our advice is to use kN and mm throughout)

The loading, however, may not be either of these two special cases. The general approach (due to Mohr) is as follows: the bending moment diagram is treated as a 'load' on the beam which will give rise to another set of 'moments' and it is these 'moments' which will give the deflections.

How to find the deflection

1   Load the beam with the BM diagram to give another set of 'moments'.

2   At any point, deflection $= \dfrac{\text{'moment'}}{EI}$

Remember,

'load' = area of BM diagram (kN m^2)

So, 'moment' has the units of kN m^3

The method may seem rather artificial – it is – but for the simply supported beam it is fairly easy to use compared with some other methods. This method is usually called the conjugate beam method since the second loading of the BM diagram is seen to act on a 'conjugate' beam. The BMs in the conjugate beam divided by EI are the deflections in the original beam.

Example 15 (Fig. 6.14a)

The smallest UB ($I = 2356$ cm^4) spans 6 m and carries 10 kN at its midpoint. Calculate the central deflection – by formula and also using the general method.

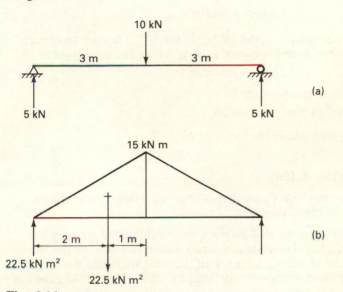

Fig. 6.14

By formula

$$y = \frac{WL^3}{48EI}$$

$$= \frac{10 \times (6000)^3}{48 \times 200 \times 2356 \times 10^4} \text{ mm}$$

$$= 9.6 \text{ mm}$$

Using the general method, we obtain the BM diagram (central BM is the 15 kN m) and treat it as a 'load' (Fig. 6.14b). The areas of the diagram are the loads.
Now

central deflection = central 'moment'/EI

118

But,

central 'moment' $= 22.5 \times 3 - 22.5 \times 1$
$$= 45.0 \text{ (kN m}^3\text{, note)}$$

Hence, working in kN and mm, we obtain:

central deflection $= \dfrac{\text{'moment'}}{EI}$

$$= \dfrac{45.0 \times 10^9}{200 \times 2356 \times 10^4} \text{ mm}$$

$$= 9.6 \text{ mm as before.}$$

Do you understand . . . the 10^4? . . . the 10^9? They are to convert cm^4 to mm^4 and m^3 to mm^3 respectively. It is easy and important to understand:

Since,   cm to mm requires 10

then, cm^4 to mm^4 requires 10^4

Can you similarly obtain the factor of 10^9?

Example 16 (Fig. 6.15a)

Determine the Universal Beam that is required to limit the midspan deflection to 1/360th of the span.

We first construct the BM diagram and calculate the areas underneath it (Fig. 6.15b). These areas are now taken as 'loads' on the beam and the left-hand reaction is simply half the total area since the diagram is symmetrical. Each of these 'loads' may be taken to act at the centroid of the area (in the centre of a rectangle but at the one-third point for a triangle).

Taking moments about the centre,

central deflection, $y = (945 \times 9 - 135 \times 7 - 270 \times 4.5$
$$-81 \times 4 - 432 \times 1.5 - 27 \times 1)/EI$$

So, $y = 5346/EI$

If the central deflection must not exceed 50 mm then we equate the central deflection to 50 mm and solve for EI, but the units must be watched. The 5346 is in kN m^3 and so requires a factor of 10^9 to make it kN mm^3.

Hence,

$$\dfrac{5346 \times 10^9}{EI} = 50$$

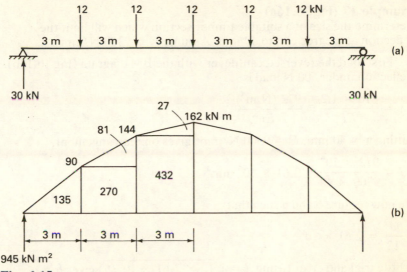

Fig. 6.15

So

$$EI = 107 \times 10^9 \text{ kN mm}^2$$

Putting $E = 200 \text{ kN/mm}^2$ gives finally,

$$I = 535 \times 10^6 \text{ mm}^4$$
$$= 53\,500 \text{ cm}^4$$

And so UB 533 × 210 × 122 kg/m is satisfactory since it has $I = 76\,207 \text{ cm}^4$. It is interesting to note that the worst stress in the beam is

$$f = \frac{M}{Z} = \frac{162 \times 10^6}{2799 \times 10^3} = 58 \text{ N/mm}^2$$

which is well below the permissible stress of 165 N/mm² for mild steel. This illustrates the fact that for longer spans the deflection is critical and not the bending moment.

What about the cantilever? How are deflections found for a cantilever? It is just as easy but there is one more additional thing to remember – the BM diagram is loaded on to the cantilever with its fixed and free ends changed around. It is too long a story to explain this satisfactorily because it would entail going into the proof of why this whole idea works anyway.

Finally: this method may be extended to other cases (for example, beams with overhanging ends) but not without a lot more thought – and it is not worth it. For the simple beam or cantilever – yes; anything else – no.

120

Example 17 (Fig. 6.16a)
Determine the size of a suitable timber section which will limit the deflection at the free end to 40 mm. Assume $E = 9300$ N/mm^2.

First load the reversed cantilever with the BM diagram (Fig. 6.16b). Deflection under 500 N load is

$$y = \frac{810 \times 1.2}{EI} = \frac{972 \ (\text{N m}^3)}{EI}$$

Putting $y = 40$ mm, $E = 9300$ N/mm^2 gives on rearrangement,

$$I = \frac{972 \times 10^9}{9300 \times 40} = 2.61 \times 10^6 \ \text{mm}^4$$

We now require b and d such that

$$\frac{bd^3}{12} = 2.61 \times 10^6$$

A little trial and error (using $d = \sqrt[3]{12 \times 2.61 \times 10^6/b}$) gives $b = 50$ mm, $d = 100$ mm, for which

$$I = \frac{bd^3}{12} = 4.17 \times 10^6 \ \text{mm}^4$$

and,

$$Z = \frac{bd^2}{6} = 83\ 300 \ \text{mm}^3$$

500 N

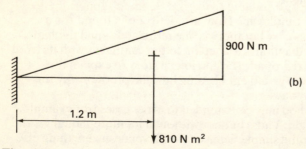

1.8 m

(a)

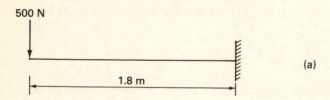

900 N m

(b)

1.2 m

810 N m^2

Fig. 6.16

Hence, the bending stress is $\quad f = \dfrac{M}{Z}$

$$= \frac{900 \times 10^3}{83\ 300}\ \text{N/mm}^2$$

$$= 10.8\ \text{N/mm}^2$$

occurring in the top and bottom edges at the support.
And the maximum shear stress is,

$$q = 1.5\,\frac{Q}{bd}$$

$$= 1.5 \times \frac{500}{50 \times 100}\ \text{N/mm}^2$$

$$= 0.15\ \text{N/mm}^2$$

occurring at the neutral axis.

6.7 Exercises

1 Explain how the reactions of a simply supported beam may be calculated. How do you deal with UDL on a beam with overhanging ends when calculating the reactions?
2 Explain how the shear force diagram may be drawn, illustrating your answer with an example of your own.
3 Explain how the bending moment diagram may be drawn. What is the basic equation for the bending moment at any point of a beam?
4 Where on a beam will the bending moment be a maximum? Where will the shear force be a maximum?
5 Distinguish between the reactions of a simple beam and those of a cantilever.
6 Draw the shear force and bending moment diagrams for the beams shown in Fig. 6.17.
7 Determine the maximum bending moment for each of the beams in Fig. 6.18 and choose a suitable Universal Beam to support them. Allow a working stress of 165 N/mm^2.
8 Timber joists 50 mm × 100 mm at 300 mm centres span 1.8 m. If their working stress must not exceed 9.3 N/mm^2 determine the safe load (kN/m^2) for the floor they support.
9 A beam spans 8 m with a UDL of 15 kN/m along its length. A single wheel load of 120 kN begins to cross the span. Determine the magnitude and postion of the maximum bending moment when the load is $x = 0, 1, 2, 3, 4$ metres from the left-hand support. What Universal Beam is required? Check the maximum shear stress. (Take care: when and where is the SF a maximum?)

122

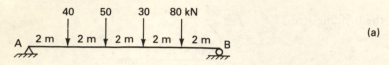

(a)

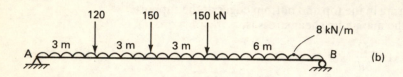

(b)

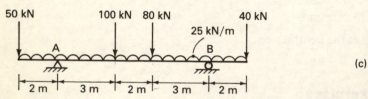

(c)

Fig. 6.17

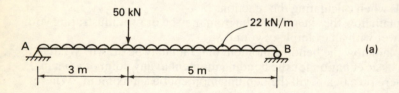

(a)

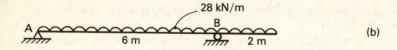

(b)

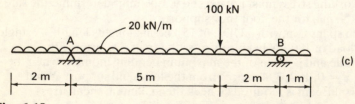

(c)

Fig. 6.18

10 Draw the SF and BM diagrams for the cantilevers shown in Fig. 6.19.

11 A load W kN acts at the centre of a span L. Prove the maximum BM is $WL/4$. Learn this!

12 A UDL of w kN/m covers a span L. Prove the maximum BM is $wL^2/8$. Learn this!

13 The Universal Beam $406 \times 140 \times 46$ kg/m ($I = 15\ 647$ cm^4) supports a central point load of 100 kN over a span of 5 m. Draw the BM diagram and hence calculate the midspan deflection. If the bending stress must not exceed 165 N/mm^2, the shear stress 100 N/mm^2 and the deflection must not exceed span/360, determine whether or not the beam is large enough.

14 Timber joists are required to support 800 N/m^2 over a span of 3 m. Assume the joists are at 300 mm centres. What UDL does one joist support? What section modulus must the joist have? Choose a suitable joist size and calculate the midspan deflection by formula. Take $E = 9700$ N/mm^2 and permissible stress $f = 9.0$N/mm^2.

15 Four 30 kN loads at 2 m centres are supported by UB $406 \times 178 \times 74$ kg/m ($I = 27\ 329$ cm^4) over a 10 m span. Calculate the midspan deflection.

16 A Universal Beam ($I = 10\ 087$ cm^4) is used as a cantilever of length 3 m. Determine the end deflections when a load of 10 kN acts at $x = 1$ m, 2 m, 3 m from the support.

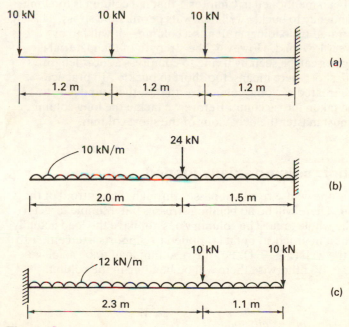

Fig. 6.19

Chapter 7

Columns

7.1 Short and long columns

It has already been mentioned in Chapter 1 that if a column is too long
then it has a tendency to buckle well below its permissible stress. This is
easily demonstrated by loading a ruler as a column – it will fail by
buckling at a very low load. However, the strength of the material is
really quite high as can be seen by trying to compress a short length
of it. The word 'short' here means 'too short to buckle'. In practice
all columns are treated as 'long' columns – that is, their tendency to
buckle must be taken into account. Before we tackle the long column,
however, we must master the behaviour of the short column.

7.2 The short column axially loaded

An axial load is one whose resultant passes through the centroid of the
column section. There will be no bending stresses but a simple direct
stress across the whole area. The column does not have the load actually
sitting on top of it in practice, but it is applied to supports fixed either to
the flanges or the web (Fig. 7.1). For the resultant load to be axial we
must have $W_1 = W_2$ otherwise there will be bending in the column.

Formula 7.1

Direct stress in a column axially loaded $f_c = \dfrac{W}{A}$

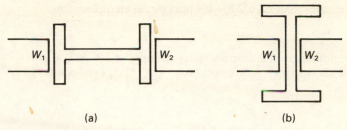

(a)   (b)

Fig. 7.1

where, $W = W_1 + W_2$ and 'c' stands for compression

Note on units: ALWAYS substitute into this formula in N and mm units. It is the only safe way. So, kN $\times 10^3$ gives N and cm$^2 \times 10^2$ gives mm^2

Example 1

The Universal Column $203 \times 203 \times 86$ kg/m ($A = 110.1$ cm^2 – see Appendix 2) supports a load of 700 kN on each flange. If the permissible stress is 155 N/mm^2, is the column safe?

From

$$f_c = \frac{W}{A}$$

we have

$$f_c = \frac{1400 \times 10^3}{110.1 \times 10^2} = 127 \text{ N/mm}^2$$

So the column is safe.

Example 2

Calculate the equal loads that can be supported either side of the web of a column with cross-sectional area 257.9 cm^2. Allow permissible stress of 155 N/mm^2.

From

$$f_c = \frac{W}{A}$$

we obtain

$$\begin{aligned} W &= f_c A \\ &= 155 \times 257.9 \times 10^2 \\ &= 4.00 \times 10^6 \text{ N} \\ &= 4000 \text{ kN} \end{aligned}$$

Since this is the total load then 2000 kN may be taken either side.

Example 3

Determine a suitable short column which will support 250 kN on each flange and 450 kN on either side of the web. Take the permissible stress to be 155 N/mm².

Clearly,

Total load = 250 + 250 + 450 + 450

= 1400 kN

Now,

$$f_c = \frac{W}{A}$$

so

$$A = \frac{W}{f_c}$$

$$= \frac{1400 \times 10^3}{155}$$

$$= 9032 \text{ mm}^2$$

$$= 90.3 \text{ cm}^2$$

The section tables (Appendix 2) show UC 254 × 254 × 73 kg/m (A = 92.9 cm²) will be suitable.

7.3 The short column eccentrically loaded

The eccentric loading arises because the loads on either side of the column are not equal (Fig. 7.1). In this case, in addition to the direct stress from $(W_1 + W_2)$ there is a bending stress due to the amount of load which is eccentric – that is, $(W_1 - W_2)$. This eccentric load, $(W_1 - W_2)$, will cause bending in the column (and $f = M/Z$ will give the stresses). We need to know how eccentric these loads are from the axis of the column. BS 449 states that the load must be assumed to act 100 mm from the face of the flange or web. Since the column has a depth D and flange width t, these distances must be {100 + D/2} or {100 + t/2} (Fig. 7.2).

Formula 7.2

Bending stress in a column eccentrically loaded $f_{bc} = \dfrac{We}{Z}$

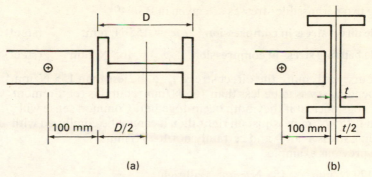

(a)   (b)

Fig. 7.2

where, $W = W_1 - W_2$ and

either, $e = 100 + \dfrac{D}{2}$ or, $e = 100 + \dfrac{t}{2}$

In addition to this bending stress $f_{bc} = M/Z$ there is also the direct stress $f_c = W/A$, and we may now combine them both.

Formula 7.3

Combined direct and bending stresses $f = \dfrac{W}{A} \pm \dfrac{M}{Z}$

Example 4 (Fig. 7.1a)
The column UC $305 \times 305 \times 158$ kg/m supports $W_1 = 1200$ kN and $W_2 = 1400$ kN. For the column, $A = 201.2$ cm², $Z_{xx} = 2368$ cm³, $D = 327.2$ mm. Calculate the worst stresses in the flanges.

We have,

direct stress, $f_c = \dfrac{W}{A} = \dfrac{2600 \times 10^3}{201.2 \times 10^2} = 129$ N/mm²

and

bending stress, $f_{bc} = \dfrac{We}{Z} = \dfrac{200 \times 10^3 \times 263.6}{2368 \times 10^3} = 22$ N/mm²

Hence,

combined stresses, $f = 129 \pm 22 = 151$, and 107 N/mm²

So the flange next to the heavier load has 151 N/mm² in it and the other flange has 107 N/mm².

We ask now, what is the permissible load for a column with axial and bending stresses? In answer we confine ourselves to steel. The

128

following two permissible stresses are given in BS 449:

Allowable direct stress in compression $p_c = 155 \text{ N/mm}^2$ (steel)

Allowable bending stress in compression $p_{bc} = 165 \text{ N/mm}^2$ (steel)

Of course we require the direct stress, f_c, to be less than 155 N/mm² and the bending stress to be less than 165 N/mm², but this requirement is clearly not sufficient if they both occur together. Common sense will show that if 155 N/mm² is just all right then it cannot be combined with any bending stress at all – and certainly not 165 N/mm².
Take the previous example:

$f_c = 129 \text{ N/mm}^2 < 155 \text{ N/mm}^2$ – all right,

$f_{bc} = 22 \text{ N/mm}^2 < 165 \text{ N/ mm}^2$ – all right.

But is the combination all right? BS 449 gives the following answer and it should be memorized.

Formula 7.4

Allowable combined direct and bending stresses $\dfrac{f_c}{p_c} + \dfrac{f_{bc}}{p_{bc}} \leqslant 1$

Applying this to the previous example gives

$$\frac{f_c}{p_c} + \frac{f_{bc}}{p_{bc}} = \frac{129}{155} + \frac{22}{165}$$
$$= 0.832 + 0.133$$
$$= 0.965$$

< 1, so the column is safe.

7.4   The Euler buckling load

The famous Swiss mathematician, Leonhard Euler (1707 – 83), calculated the load at which a column would buckle if it were axially loaded and pinned at its ends (Fig. 7.3). Although the mathematics is very interesting we shall not go into that now but simply give the final result that Euler proved.

Formula 7.5

The Euler buckling load for an axially loaded pin ended column $P_E = \pi^2 EI/L^2$

This is certainly an intriguing result. As we ponder it we can see that the buckling load is quite independent of the strength of the material (as measured by its permissible stress). The length of the column (L) is

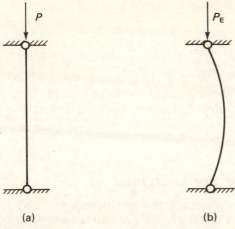

(a)   (b)

Fig. 7.3

clearly important. As the length increases so the buckling load comes down, quite rapidly. Indeed, if the length is doubled the buckling load is reduced to one-quarter of its previous value.

The presence of the second moment of area I is interesting too: since the column will always buckle about its weaker axis this means that I must refer to this axis. A piece of paper standing as a column along one of its edges is extremely flimsy – it almost buckles under its own self-weight. This is because the value of I for the axis about which it buckles is exceedingly small. However, if the same piece of paper is rolled to make a cylindrical column it can support a greater weight by far – a small book – without buckling. Try it and see. Although the cross-sectional areas of the two columns are the same, the value of I for the paper roll is far higher.

Example 5

Calculate the Euler buckling load for the smallest Universal Column of length 4.15 m. Compare this load with the load it can support as a short column. Take $I_{yy} = 403$ cm^4, $A = 29.8$ cm^2.

We have

$$\text{buckling load, } P_E = \frac{\pi^2 EI}{L^2}$$

So

$$P_E = \frac{\pi^2 \times 200 \times 403 \times 10^4}{4150^2} = 462 \text{ kN}$$

130

As a short column, allowing 155 N/mm^2, it can take

$W = f_c A$
$\quad = 155 \times 29.8 \times 10^2/10^3$
$\quad = 462$ kN

So it is clear that while the column is quite safe with a load of 462 kN if it is short, it will nevertheless collapse by buckling under this load if it is 4.15 m high.

7.5 Factors affecting the buckling load

Two factors have already been mentioned in the previous section, and are obvious from the formula for P_E – the length, L, and the second moment of area, I. Another factor is the Young's modulus, E, which is normally that for steel or concrete. One of the most important factors however, is the degree of fixity at the ends of the column. The practical column is not pinned at its ends. It may be rigidly fixed or partially fixed and this needs to be taken into account. Take your ruler again and compare the buckling loads for the two end conditions – pinned and fixed. One is four times the other.

It is clear that some means is required by which the allowable load for a long column may be calculated which will take into account the tendency of the column to buckle. It will no longer be simply a matter of taking the cross-sectional area of the column and multiplying it by the permissible stress – account must also be taken of the length, end fixity and second moment of area.

7.6 The slenderness ratio

A column's tendency to buckle may be measured by its slenderness ratio defined in BS 449 as follows.

Definition 7.1

SLENDERNESS RATIO, $\quad s = \dfrac{\text{end fixity} \times \text{length}}{\text{least radius of gyration}}$

where the end fixity is given as follows:

end condition	pinned	intermediate	fixed
end fixity	1.0	0.85	0.7

Note 1 The slenderness ratio has no units – so be sure to substitute the length and least radius of gyration in the same units (mm is as good a choice as any).

Note 2 The term 'end fixity' is a factor designed to take account of how the ends of a column are restrained. There are other factors for other end conditions but for simplicity we confine attention to the three above. For practical work the most popular is 0.85.

Note 3 The least radius of gyration is serving the same function as the second moment of area in the Euler formula. Remember always to look up the radius of gyration for the $Y–Y$ axis, as this is the axis about which buckling will occur.

Example 6

Compare the slenderness ratios of the largest and smallest Universal Columns if they are fixed at their ends and are each 3 m high. Their least radii of gyration are respectively 11.0 cm and 3.68 cm.

Working in mm we have,

$$s_1 = \frac{0.7 \times 3000}{110} = 19 \qquad \text{for the largest UC,}$$

$$s_2 = \frac{0.7 \times 3000}{36.8} = 57 \qquad \text{for the smallest UC.}$$

Example 7

A column 4.1 m high with partial end fixity has radii of gyration about the $X–X$ and $Y–Y$ axes of 14.8 and 8.25 cm respectively.
Calculate the slenderness ratio.

Remembering that we must use the least radius of gyration, we find

$$s = \frac{0.85 \times 4100}{82.5} = 42$$

Example 8

Determine the slenderness ratio of a steel plate 2 m long effectively pinned at its ends and of rectangular section 200 mm × 25 mm.

We know,

$$r = \sqrt{\frac{I}{A}}$$

where (working about the weaker axis of bending),

$$I = \frac{bd^3}{12} = \frac{200 \times 25^3}{12} = 260 \times 10^3 \text{ mm}^4$$

and

$$A = 200 \times 25 = 5000 \text{ mm}^2$$

hence,

$$r = \sqrt{\frac{260 \times 10^3}{5000}} = 7.2 \text{ mm}$$

Therefore the slenderness ratio, s, is given by,

$$s = \frac{1.0 \times 2000}{7.2} = 278$$

This is so much higher than the other values previously calculated because of course the plate is so slender compared with a column.

7.7 The permissible load for a long column

We have now arrived at the final hurdle. How should the strength of a long column be calculated? How can its slenderness be taken into account? Many ways have been suggested but we will choose just one of them. The basis of it is to lower the permissible direct stress, p_c, according to the value of the slenderness ratio: the greater the slenderness ratio the lower the permissible stress. Table 7.1 lists the values given in BS 449.

Table 7.1

Slenderness ratio (s)	Allowable direct stress in compression (p_c) N/mm^2
0	155
10	151
20	147
30	143
40	139
50	133
60	126
70	115
80	104
90	91
100	79

Note: The allowable bending stress (p_{bc}) remains at 165 N/mm^2 for this range of slenderness ratio when applied to the Universal Column range.

The condition for the column to be safe under combined bending and direct stresses remains as before:

$$\frac{f_c}{p_c} + \frac{f_{bc}}{p_{bc}} \leqslant 1$$

but p_c is lowered from 155 N/mm^2 according to the slenderness ratio.

Example 9 (Fig. 7.4)

The Universal Column $254 \times 254 \times 132$ kg/m supports loads of 420 kN and 550 kN on either flange. The height of the column is 3.8 m. The section properties are $Z_{xx} = 1622$ cm^3, $Z_{yy} = 570.4$ cm^3, $r_{xx} = 11.6$ cm, $r_{yy} = 6.66$ cm, $A = 167.7$ cm^2, $D = 276.4$ mm. Can the column safely support the loads? Allow partial end fixity.

This is an important problem and it is essential to understand each step.

For safety we require

$$\frac{f_c}{p_c} + \frac{f_{bc}}{p_{bc}} \leqslant 1$$

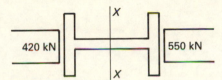

Fig. 7.4

Now,

$$f_c = \frac{W}{A} = \frac{970 \times 10^3}{167.7 \times 10^2} = 57.8 \text{ N/mm}^2$$

and,

$$f_{bc} = \frac{We}{Z} = \frac{130 \times 10^3 \times 238.2}{1622 \times 10^3} = 19.1 \text{ N/mm}^2$$

Quick questions: Can you explain where each one of these numbers come from?
Why 10^2: Why each 10^3? Why 1622, not 570.4? Why 970? Why 130? Why 238.2?

Now we turn to the permissible stresses p_c and p_{bc}. Since p_c depends on the slenderness ratio, we have

$$s = \frac{0.85 \times 3800}{66.6} = 48$$

and so,

$$p_c = 134 \text{ N/mm}^2 \qquad \text{(from Table 7.1)}$$

Finally,

$$p_{bc} = 165 \text{ N/mm}^2 \qquad \text{(as always)}$$

Hence,

$$\frac{f_c}{p_c} + \frac{f_{bc}}{p_{bc}} = \frac{57.8}{134} + \frac{19.1}{165}$$

$$= 0.431 + 0.116$$

$$= 0.547$$

$$< 1, \text{ so the column is quite safe.}$$

Example 10 (Fig. 7.5)

The column is the same as in the previous example. The web thickness is $t = 15.6$ mm. Can the column support the 630 kN off the web?

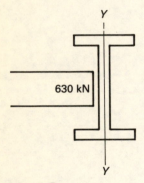

Fig. 7.5

For safety

$$\frac{f_c}{p_c} + \frac{f_{bc}}{p_{bc}} \leqslant 1$$

where

$$f_c = \frac{W}{A} = \frac{630 \times 10^3}{167.7 \times 10^2} = 37.6 \text{ N/mm}^2$$

and

$$f_{bc} = \frac{We}{Z} = \frac{630 \times 10^3 \times 107.8}{570.4 \times 10^3} = 119.1 \text{ N/mm}^2$$

As before,

$$p_c = 134 \text{ N/mm}^2, \qquad p_{bc} = 165 \text{ N/mm}^2$$

So,

$$\frac{f_c}{p_c} + \frac{f_{bc}}{p_{bc}} = \frac{37.6}{134} + \frac{119.1}{165}$$
$$= 0.280 + 0.722$$
$$= 1.002$$
$$> 1$$

so this column is overstressed and a larger one is required.

Example 11 (Fig. 7.6)

Check the safety of this UC 305 × 305 × 283 kg/m, 4.2 m high and rigidly fixed at its ends.

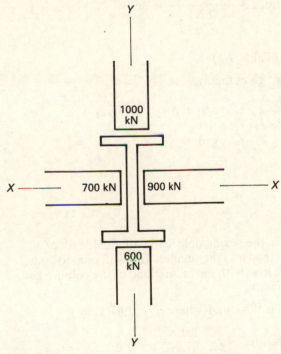

Fig. 7.6

In this case we have bending about both X–X and Y–Y axes but if we take them separately all will be well.

For safety,

$$\left(\frac{f_c}{p_c}\right) + \left(\frac{f_{bc}}{p_{bc}}\right)_{xx} + \left(\frac{f_{bc}}{p_{bc}}\right)_{yy} \leqslant 1$$

where

$$f_c = \frac{W}{A} = \frac{3200 \times 10^3}{360.4 \times 10^2} = 88.8 \text{ N/mm}^2$$

X–X axis:

$$f_{bc} = \frac{We}{Z} = \frac{400 \times 10^3 \times 282.65}{4314 \times 10^3} = 26.2 \text{ N/mm}^2$$

Y–Y axis:

$$f_{bc} = \frac{We}{Z} = \frac{200 \times 10^3 \times 113.45}{1525 \times 10^3} = 14.9 \text{ N/mm}^2$$

Now, the slenderness ratio, $s = \dfrac{0.7 \times 4200}{82.5} = 36$

Hence,

$$p_c = 141 \text{ N/mm}^2 \qquad \text{(Table 7.1)}$$

Putting $p_{bc} = 165 \text{ N/mm}^2$ gives finally,

$$\left(\frac{f_c}{p_c}\right) + \left(\frac{f_{bc}}{p_{bc}}\right)_{xx} + \left(\frac{f_{bc}}{p_{bc}}\right)_{yy} = 0.630 + 0.159 + 0.090$$

$$= 0.879$$
$$< 1$$

so the column is safe.

Example 12

Determine the variation in the permissible load with the length of a column. Assume an axial load and the smallest UC with pinned ends. Compare this permissible load with the failure load of the column and hence deduce the safety factor.

The permissible load is $W = p_c A$ (where $A = 2980 \text{ mm}^2$)

Now, the slenderness ratio, $\qquad s = \dfrac{1.0 \times L}{0.0368}$

where L is the length of the column in metres and of course the least radius of gyration is 0.0368 m. Table 7.1 will give the values of p_c hence we may calculate the permissible load W for various values of L. This is shown in Fig. 7.7a.

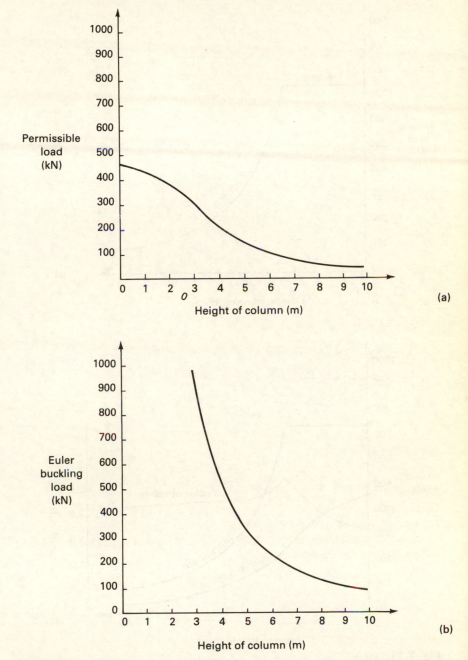

Fig. 7.7

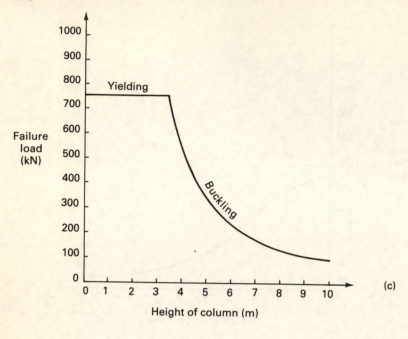

Failure load (kN)

Yielding

Buckling

Height of column (m)

(c)

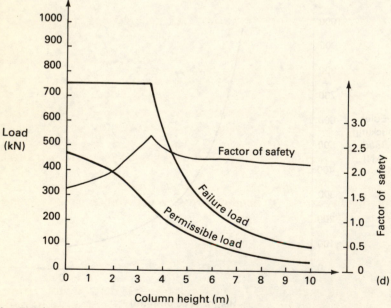

Load (kN)

Factor of safety

Factor of safety

Failure load

Permissible load

Column height (m)

(d)

Fig. 7.7 (*cont.*)

The failure load, due to buckling, is the Euler load, P_E, where,

$$P_E = \frac{\pi^2 EI}{L^2}$$

Substituting $E = 210$ kN/mm^2 and $I = 4.03 \times 10^6$ mm^4 will give values of P_E for various values of L. This graph is shown in Fig. 7.7b.

A little thought will make it clear that for short columns the failure will not be by buckling but by yielding of the material. Taking the yield stress of mild steel as $f_y = 250$ N/mm^2, we find the failure load for a short column is

$$W = f_y A$$
$$= 250 \times 2980/10^3$$
$$= 745 \text{ kN}$$

This may be combined with the buckling curve to give the failure load over the whole range (Fig. 7.7c). Comparing this with the permissible load gives Fig. 7.7d. A final definition of safety factor completes the problem.

Definition 7.2

$$\text{SAFETY FACTOR} = \frac{\text{failure load}}{\text{permissible load}}$$

Hence we may obtain the safety factor curve given also in Fig. 7.7d.

7.8 Exercises

Note: Universal Column tables are to be found in Appendix 2.

 1 Quote the formula for direct stress and explain the symbols you use.
 2 Quote the formula for bending stress and explain the symbols you use.
 3 Quote the formula which combines direct and bending stresses and explain it.
 4 What are the permissible stresses in mild steel for (*a*) direct compressive stress, (*b*) bending stress? What symbols would you use for the stresses?
 5 State the relationship which must exist between the actual and permissible stresses in a column for it to be safe.
 6 Quote the formula for the Euler buckling load.
 7 Define slenderness ratio.
 8 Explain the significance of buckling when applied to columns. How is buckling taken into account when calculating the permissible load on a column?
 9 State the eccentricity of a load applied to (*a*) the flange (*b*) the web of a Universal Column.

140

10 Define safety factor. Sketch typical curves for failure load and permissible load against height for a column showing how the safety factor might vary.

11 Choose a suitable Universal Column to support 1000 kN off each flange. Assume a short column. What column is required if the loads come either side of the web?

12 Plot a graph to show the permissible axial load for the whole Universal Column range. The columns should be specified on the x-axis and the permissible load on the y-axis. Assume short columns.

13 The UC 203 × 203 × 46 kg/m supports 210 kN off each flange and 180 kN off each side of the web. It is 3 m high and it is partially fixed at its ends. Check whether the column is suitable.

14 Determine the permissible axial load of UC 305 × 305 × 158 kg/m if it is 3.8 m high and has fixed ends.

15 Select a suitable Universal Column, of length 3.2 m, pinned at its ends, which will safely support an axial load of 2000 kN.

16 The UC 254 × 254 × 167 kg/m is pinned at its supports and is 3.5 m high. Determine whether or not it can support an eccentric loading of 800 kN and 1200 kN (*a*) on the flanges, (*b*) on either side of the web.

17 The smallest UC is fixed at its ends which are 3.0 m apart. Can it support 100 kN (*a*) off a flange, (*b*) off the web? Draw the stress distribution across the column cross-section for both these cases.

18 The UC 254 × 254 × 73 kg/m carries 227 kN on one flange and 93 kN on one side of the web. Its height is 3.0 m. Check its safety for the three end conditions (*a*) pinned, (*b*) fixed, (*c*) partial fixity. What are the stresses at the four corners of the column?
(*Hint:* $f = f_c \pm (f_b)_{xx} \pm (f_b)_{yy}$)

19 The UC 305 × 305 × 97 kg/m is rigidly fixed at its ends and is 4.0 m high. It is loaded initially with 800 kN on each flange. Calculate $G = (f_c/p_c) + (f_{bc}/p_{bc})$. Show how G varies as the load (W) on one of the flanges varies between 0 – 1000 kN. Draw a graph of the result marking on your graph the safe range of W.

20 Part of the web of a beam is subjected to a vertical force and it needs to be checked for buckling. The part of the web taking the force may be taken as a vertical plate 19.2 mm thick, 246 mm high, 362 mm long and rigidly fixed along its top and bottom edge. Calculate the slenderness ratio and p_c and hence determine the load the web may safely support.

Chapter 8

Trusses

8.1 The merits of a truss

The chief advantage of a truss is its high strength-to-weight ratio. Each member of the truss is (at least theoretically) in simple tension or compression and therefore all the material may be put to good use. In contrast, a section in bending must of necessity have a zone of low stress near the neutral axis. The truss is rather like a beam with large parts of it removed leaving only the members behind.

8.2 Determinate and indeterminate trusses

Normally the members of a truss form a network of triangles. This is essential for the truss to be rigid. If, for example, the members formed squares then the whole truss would collapse since it would not be rigid. Structures can indeed be made with members forming squares, it is true, but these are not trusses since the members would need to be rigidly connected at the joints whereas the truss members are pinned together.

How many members are required for a truss with j joints? For each new joint two members are required so, roughly speaking, there are twice as many members as joints. For any particular case we can see we need three fewer members than twice the number of joints.

Formula 8.1
For a truss, $m = 2j - 3$ where j is the number of joints and m is the number of members to make a rigid truss.

142

Such a truss – which has just the right number of members to make it rigid – is capable of being analysed by the simple principles of equilibrium alone. It is therefore called a determinate truss as the member forces can be determined by elementary resolving and taking moments. Any fewer members than $m = 2j - 3$, and the truss is a mechanism and can support no load at all. Any more members than $m = 2j - 3$, and the truss is called indeterminate because it is impossible to determine the forces by equilibrium alone. To solve an indeterminate truss requires deflections to be taken into account and it is a much more complicated affair – we shall not tackle it in this book. Figures 8.1a, b and c give the three cases. The additional members above $m = 2j - 3$ are called redundant members. These must not be confused with members with no force in them, which are by no means necessarily redundant.

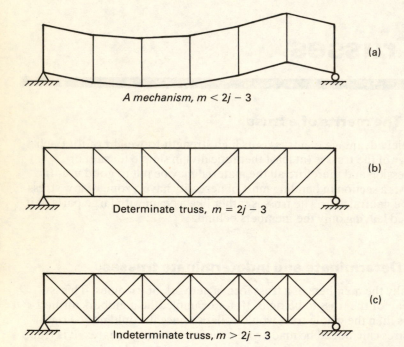

A mechanism, $m < 2j - 3$ (a)

Determinate truss, $m = 2j - 3$ (b)

Indeterminate truss, $m > 2j - 3$ (c)

Fig. 8.1

8.3 The method of joints

We come now to the first method of finding the member forces in a truss. This is how we set about it:

Method of joints Start at the joint with just two members.
1 Resolve perpendicular to one member

2 Resolve again in another convenient direction.
This will give the forces in the two members. Now move to the next joint
with just two unknown member forces and repeat the procedure.

Example 1 (Fig. 8.2)

Joint A	Resolve vertically:	$F_1 = 150$ kN
	Resolve horizontally:	$F_2 = 0$
Joint B	Resolve vertically	$F_3 \cos 45 = F_1$
	hence	$F_3 = 212$ kN
	Resolve horizontally:	$F_4 = F_3 \cos 45$
	hence	$F_4 = 150$ kN
Joint C	Resolve vertically:	$F_5 + 100 = F_3 \cos 45$
	hence	$F_5 = 50$ kN
	Resolve horizontally:	$F_6 = F_2 + F_3 \cos 45$
	hence	$F_6 = 150$ kN
Joint D	Resolve vertically:	$F_7 \cos 45 = F_5$
	hence	$F_7 = 71$ kN
	Resolve horizontally:	$F_8 = F_7 \cos 45 + F_4$
	hence	$F_8 = 200$ kN

Since the frame is symmetrical there remains only F_9 to find.

| Joint F | Resolve vertically: | $F_9 = 0$ |

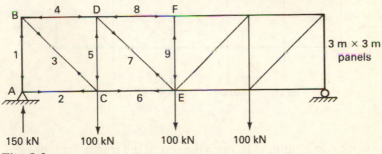

Fig. 8.2

Note 1 Arrows, denoting the forces of the members on the joints, must
first be put on the truss. With a normal loading it is possible to get these
right by remembering the upper members are in compression and the
lower members in tension; the verticals compression, the diagonals
tension. Remember that tension is denoted by inward arrows. Alternatively, put them all in tension and you will then find all positive
answers will be tension and negative answers compression.

Note 2 Only consider the forces acting at the joint you are looking at.
At joint D, for example, only forces 4, 5, 7, 8 are acting – *not* force 3 or
6 or the 150 kN reaction, or even the 100 kN load.

8.4 A simplified method of joints for parallel boom trusses

Simplified method of joints

1 Using the *components* of the diagonal member forces resolve at each joint as before.
2 Calculate the forces in the diagonals.

Example 2 (Fig. 8.3)

First we find the reactions then begin at joint A.

Joint A The reaction of 100 kN up must be resisted by 100 kN down. Hence force in AB is 100 kN compression.
In the horizontal direction no force comes on to it so no force is required in member AC.

Joint B The force of 100 kN up from member AB must be resisted by 100 kN down in the diagonal. This is the vertical component of the diagonal. Since the diagonal is at 45°, the horizontal component must also be 100 kN. Resolving horizontally the diagonal is pulling joint B to the right with a force of 100 kN. So member BD must push it to the left with a force of 100 kN.

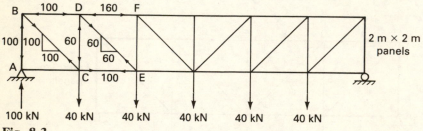

Fig. 8.3

The work now continues through the rest of the frame. It involves only simple addition or subtraction and the balancing of the forces in one direction with forces in the opposite direction.

Q1: Do you agree with the forces at joint C, (a) vertically?
(b) horizontally?
Q2: Do you agree with the forces at joint D?
Q3: What are the unknown forces at joint E?
Finally the actual diagonal forces are $100\sqrt{2}$, $60\sqrt{2}$, and so on.

This simplified method is very quick and simple and worth mastering. It can, however, only be applied with parallel top and bottom booms. The general approach can be applied whatever the frame is like. If the panels are not square then the horizontal component is not the same as the vertical component.

Diagonals at X degrees to the horizontal
If the vertical component is V then,

horizontal component $= V/\tan X$

diagonal force $\qquad = V/\sin X$

The principle remains the same as before – balancing the forces vertically and horizontally at each joint.

8.5 The method of sections

This second method has a beauty all of its own. We have met it before in the chapter on internal forces. Often we need to know the force in just one member – the member with the greatest force in it, for example – and the method of sections will yield the force in a particular member without the labour of working out the forces in the rest of the truss.

Method of sections
1 Cut the truss into two parts through the member you want and two other 'odd' members.
2 For one half of the truss only,
either: take moments about the point where the two 'odd' members intersect;
or: resolve perpendicular to the two 'odd' members.

Example 3 (Fig. 8.4)
Calculate the member force X.

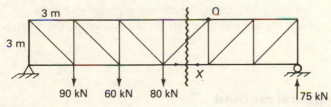

Fig. 8.4

Let us cut through X and the two members above it and consider the equilibrium of the right half of the truss. Taking moments about Q (this is the point where the two 'odd' members intersect),

$\qquad X \times 3 = 75 \times 6$

Therefore,

$\qquad X = 150$ kN tension.

Q: Can it be this simple?   *A: Yes!*

Example 4 (Fig. 8.5)

Calculate the member force Y.

We cut through Y and two other members as shown. Since the two 'odd' members are parallel they do not intersect and we must therefore resolve perpendicular to them.
Resolving vertically for the right half,

$Y \cos 30 = 25$

Hence,

$Y = 28.9$ kN tension.

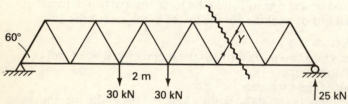

Fig. 8.5

Note 1 You may consider either half of the frame – but once you have decided you must forget about the other half.

Note 2 The member force may be marked tension or compression – it does not matter – but you must decide which and put arrows on the member accordingly. If the answer is negative then you made the wrong choice.

8.6 The graphical method

This method is essentially an extension of the triangle of forces.

The graphical method
1 Moving clockwise around the outside of the frame, draw the force diagram for the external forces.
2 Taking each joint in turn, draw the force diagram for the unknown forces at each joint.

It is best to explain these steps by means of an example. We note that, like the method of joints (but unlike the method of sections), work proceeds from one end of the frame to the other.

Example 5 (Fig. 8.6)

Determine the member forces.

First the reactions are calculated and the spaces between the forces and members lettered (Fig. 8.6a). Moving clockwise round the frame (keeping watch on both Figs. 8.6a and 8.6b),

AB is 150 kN up — so draw *ab* 15 mm up

BC is 200 kN down — so draw *bc* 20 mm down

CD is 200 kN down — so draw *cd* 20 mm down

DA is 250 kN up — so draw *da* 25 mm up

Note that this part of the diagram closes back on to itself.

Now we begin to find the member forces. Starting first with the joint with just two members at the left support, we have:

DE is a horizontal member — draw *de* horizontally

AE is a vertical member — draw *ae* vertically

de and *ae* will intersect at the point *e*.

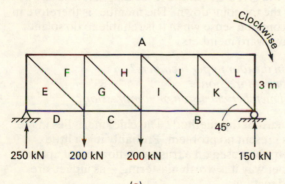

(a)

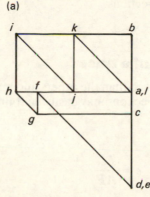

(b)

Fig. 8.6

Similarly,

> EF is at 45°   – draw *ef* at 45°
> AF is horizontal – draw *af* horizontally
> hence the point *f*.

Again,

> FG (vertical) and CG (horizontal)

will give,

> *fg* (vertical) and *cg* (horizontal), and hence the point *g*.
> Can you arrive at the points *h*, *i*, *j*, *k* and *l*?

When the force diagram (*a*, *b*, *c*, . . .) is complete, the force in a member is found from the length of the appropriate line. For example, member EF has a 350 kN force in it since *ef* = 35 mm. To find whether it is tension or compression: take one end of the member and go round it clockwise noting the order of the spaces as you cross the member. For the upper end of EF this will give FE. On the force diagram *fe* is down so the member is pulling the top joint down. The member is therefore in tension. It is best to use common sense when it is reliable to do so and use this method for the awkward members.

Q1: What are the forces in the four upper members?
Q2: Are they in tension or compression?
Q3: Is the central vertical member in tension or compression?

Finally, what is the merit of this method? The chief merit lies in the fact that awkward angles present no problem. Probably it is a little slower than by calculation when used on a truss with horizontal upper and lower members. Either way it is worth mastering – as indeed are the methods of joints and sections.

8.7   The deflection of a truss

This section is more advanced than the others and may be missed at the first reading. The truss deflections may easily be found from the member extensions.

Formula 8.2

Extension of a member $e = \dfrac{FL}{AE}$

where   *F* is the tension, *L* is the length,
          A is the area, *E* is the Young's modulus.

Formula 8.3
Deflection of a truss $d = \Sigma ef$

where e is the set of member extensions,
             f is the set of member forces due to a unit load acting at the
                joint for which the deflection is required,

and Σef means the sum of the products $\{ef\}$ for each member.

Example 6 (Fig. 8.7)

The truss shown in Fig. 8.7a is composed of identical members each of
length 3 m, and cross-sectional area 1000 mm² and Young's modulus
200 kN/mm². Calculate the deflection under the load.

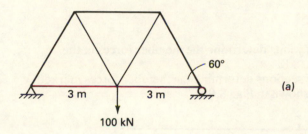

(a)

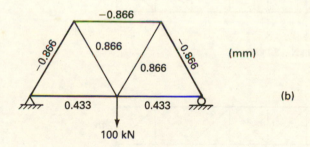

(mm)

(b)

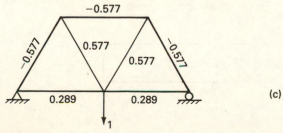

(c)

Fig. 8.7

150

We first calculate the member forces (F) and hence calculate the member extensions from Formula 8.2. These member extensions (e) are shown in Fig. 8.7b in millimetres with the sign convention extension ($+$), contraction ($-$). Since the central deflection is required we remove the loading and place a unit load at the centre. The unit load member forces (f) are shown in Figure 8.7c.

Taking each member in turn we add up the products $\{ef\}$ to give the required deflection according to Formula 8.3.

Hence,

central deflection, $d = \Sigma ef$

$$= 2(0.433 \times 0.289) + 5(0.866 \times 0.577)$$

$$= 2.7 \text{ mm}$$

8.8 Exercises

1 Using the method of joints determine the member forces in the trusses shown in Fig. 8.8.
2 Using the method of sections determine the member forces marked X and Y in the trusses shown in Fig. 8.9.

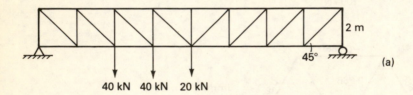

40 kN 40 kN 20 kN

45° (a)

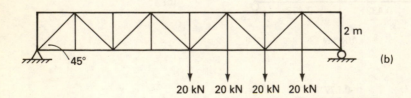

45°

20 kN 20 kN 20 kN 20 kN

2 m (b)

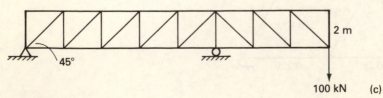

45°

2 m

100 kN (c)

Fig. 8.8

151

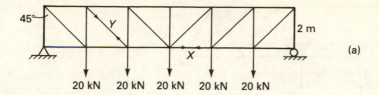

(a)

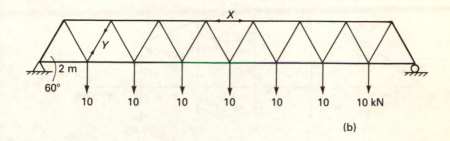

(b)

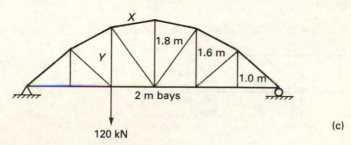

(c)

Fig. 8.9

3 Using the graphical method determine the member forces in the trusses shown in Fig. 8.10.

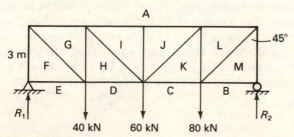

Fig. 8.10(a)

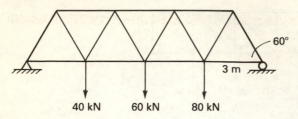

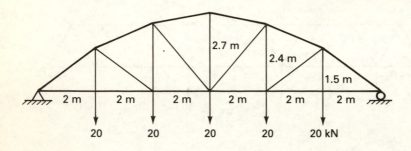

Fig. 8.10(b) and (c)

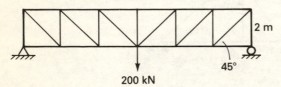

Fig. 8.11

4 Calculate the central deflection of the truss shown in Fig. 8.11. Take
$L/AE = 0.01$ mm/kN for each member.

5 A single load of 1 kN rolls across a Warren girder spanning 12 m
composed of members each 2 m long. Calculate the compressive
force in the central horizontal member when the load is placed at each
of the lower chord joints. Plot a graph to show the variation in the
force in this member. Plot a similar graph to show how the force in
one of the central diagonals varies. These graphs are called influence
lines and are used when a group of loads rolls over. Can you see how
you could use these influence lines to calculate the worst force in the
members when three loads, 100 kN each and 2 m apart, pass over?

Chapter 9

Arches

9.1 The advantage of the arch

It has been pointed out in an earlier chapter that the bending moments in an arch rib have been considerably reduced by the action of the horizontal thrust at the abutments. Without such horizontal thrust the bending moments would be just the same as those in a beam of the same span. Of course, in exchange for the bending moments the arch rib has received a fairly high axial thrust. In conclusion, the arch rib experiences a high thrust with small moments for which a much smaller section is required than that for a corresponding beam. Arch spans therefore can be much longer than the spans for beams.

9.2 Determinate and indeterminate arches

The basic principles of equilibrium will provide only three independent equations. This means that if the problem has only three unknowns then they may be found by simple resolving and taking moments. However, with four or more unknowns the structure is said to be indeterminate since it is impossible to find the four or more equations by resolving or taking moments.

Look at Fig. 9.1a, which shows an arch fixed to rigid abutments. At each support there are three unknowns, making a total of six unknowns in all. Statics (that is, resolving and taking moments) will give three equations and hence there are three further equations to be found from somewhere else. The structure has therefore three redundant reactions.

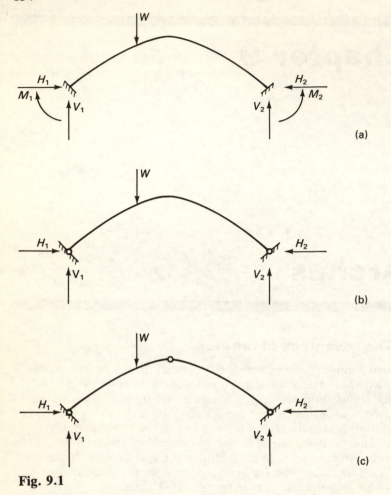

Fig. 9.1

Look at Fig. 9.1b. Here the arch is pinned to the supports. Two unknowns at each support make a total of four in all. Statics will give three equations which leaves one further equation to find. The structure has therefore one redundant reaction.

Now consider Fig. 9.1c. There are four unknown reactions. At first it may look as though there is a redundant reaction here also. However, the pinned joint in the crown means that there is no moment there and so with the three equations of statics we may add one more which will give the four required to solve for the reactions.

This is the only type of arch which we shall consider – the statically determinate three-pin arch. It is an interesting structure to analyse, but before we get to that we need to look at the geometry of the arch.

9.3   The equation of the arch shape

An arch is shown in Fig. 9.2 referred to x and y axes. Since the forces
and moments in an arch depend upon its shape it is necessary to know its
equation. By far the easiest and commonest arch curve is the parabola.
Sometimes the shape is an arc of a circle but this is far harder to solve
and less common.

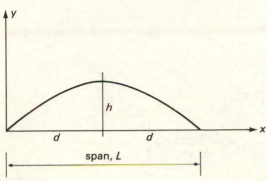

Fig. 9.2

Formula 9.1

Equation of a parabolic arch $y = \dfrac{4h}{L^2} x(L - x)$

It is also important to know the angle, θ, the arch makes with the
horizontal. This is easily found for a parabola and the result is as
follows:

Formula 9.2

Slope of a parabolic arch   Slope $= \dfrac{4h}{L^2} (L - 2x)$

Angle to horizontal, $\theta = $ arc tan (slope).

9.4   The reactions of a three-pin arch

The reactions do not depend on the arch profile but only on the
positions of the pins. The general approach is as follows:

How to obtain the reactions of a three-pin arch
1 Take moments about the right support.
2 Take moments about the crown for the left half of the arch. This will
   give the reactions at the left support.

3 Resolve vertically.
4 Resolve horizontally. This will give the reactions at the right support.

Example 1 (Fig. 9.3)
Calculate the reactions for this arch.

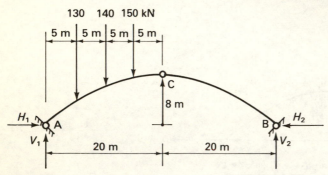

Fig. 9.3

Taking moments about B, we have

clockwise moments = anticlockwise moments

So,

$$V_1 \times 40 = 130 \times 35 + 140 \times 30 + 150 \times 25$$

Hence,

$$V_1 = 312.5 \text{ kN}$$

Taking moments about C for the left half of the arch, we have

anticlockwise moments = clockwise moments

So

$$H_1 \times 8 + 130 \times 15 + 140 \times 10 + 150 \times 5 = V_1 \times 20$$

Substituting for V_1 then gives,

$$H_1 = 268.8 \text{ kN}$$

Consider the equilibrium of the whole arch again:

Resolving vertically,

forces up = forces down

Therefore,

$$V_1 + V_2 = 130 + 140 + 150$$

Hence,

$V_2 = 107.5$ kN

Resolving horizontally,

$H_2 = H_1$

So,

$H_2 = 268.8$ kN

Example 2 (Fig. 9.4)

Determine the reactions of this arch.

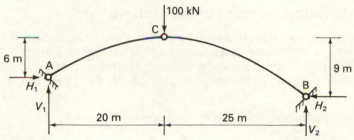

Fig. 9.4

    Here we see the supports are at different levels but the simple principles of equilibrium will carry us through.
Taking moments about B,

clockwise moments = anticlockwise moments

So

$V_1 \times 45 + H_1 \times 3 = 100 \times 25$

Therefore,

$45V_1 + 3H_1 = 2500$

This cannot be solved immediately because it has two unknowns. We therefore take moments about C for the left half of the structure;

clockwise moments = anticlockwise moments

Hence,

$V_1 \times 20 = H_1 \times 6$

or

$10V_1 - 3H_1 = 0$

158

These two equations may now be solved for V_1 and H_1. Adding them we obtain,

$$55V_1 = 2500$$

So,

$$V_1 = 45.5 \text{ kN}$$

Hence,

$$H_1 = 151.5 \text{ kN}$$

Resolving vertically and horizontally gives $V_2 = 54.5$ kN, $H_2 = 151.5$ kN.

9.5 The bending moment in the arch rib

Bending moments in an arch, though much less than those in a beam of the same span, are nevertheless significant and need to be calculated.

Consider Fig. 9.5, which shows part of an arch rib to the left of a point X at which the bending moment, M, is required. Taking moments about X,

anticlockwise moments = clockwise moments

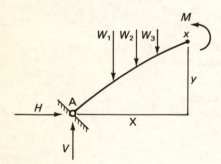

Fig. 9.5

So

$$M + Hy = Vx - \text{moment of loads}$$

But $\{Vx - \text{moment of loads}\}$ is the normal beam moment. Hence,

$$M + Hy = \text{beam moment}$$

Therefore,

$$M = \text{beam moment} - Hy$$

This shows clearly how the arch moments are much less than the beam moments. Let us summarise it again and memorise it.

How to find the BM in an arch

Arch moment = beam moment − *Hy*

Example 3 (Fig. 9.6a)

Determine the bending moments at 10 m centres across this arch.

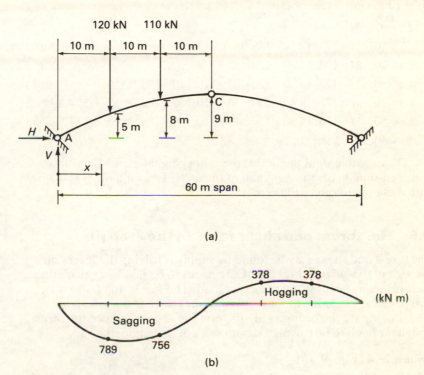

(a)

(b)

Fig. 9.6

First we must find the reactions.
Taking moments about B for the whole arch,

$$V \times 60 = 120 \times 50 + 110 \times 40$$

So,

$$V = 173.3 \text{ kN}$$

Taking moments about *C* for the left half,

anticlockwise moments = clockwise moments

Hence,

$$H \times 9 + 120 \times 20 + 110 \times 10 = 173.3 \times 30$$

So,

$H = 188.9$ kN

To find the arch moments across the span, we have

arch moment = beam moment − Hy

So at 10 m centres we obtain:

at $x=0$, $M=0$
at $x=10$, $M=173.3\times10-188.9\times5$ $=788.9$ kNm
at $x=20$, $M=173.3\times20-120\times10-188.9\times8$ $=755.6$ kNm
at $x=30$, $M=173.3\times30-120\times20-110\times10-188.9\times9=0$ *(why?)*
at $x=40$, $M=173.3\times40-120\times30-110\times20-188.9\times8=-377.8$ kNm
at $x=50$, $M=173.3\times50-120\times40-110\times30-188.9\times5=-377.8$ kNm
at $x=60$, $M=0$, of course.

The negative signs mean that it is a hogging moment. Does this seem reasonable for the right half of the arch? Drawing the BM diagram with these results gives Fig. 9.6b.

9.6 The thrust and shear force in the arch rib

The thrust and shear may be found by simply resolving the forces on the arch in the correct direction. Care needs to be taken in calculating their values at the position of a point load. The fact is that there is an abrupt change in value of the thrust and shear either side of a point load – compare the ordinary SF diagram for a beam. Hence the thrust and shear need to be calculated both sides of a point load.

Example 4 (Fig. 9.7a)
Determine the variation in thrust at 10 m centres across the arch.

Taking moments about B gives

$V = 300$ kN

Taking moments about C,

$H \times 8 + (15 \times 20) \times 10 = 300 \times 20$

So,

$H = 375$ kN

At the quarter point $x = 10$ m

So, from

$$\text{slope} = \frac{4h}{L^2} (L - 2x)$$

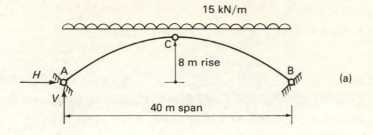

(a)

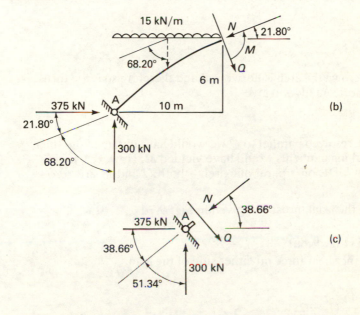

(b)

(c)

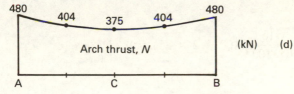

Arch thrust, N

(kN) (d)

A   C   B

Fig. 9.7

we have

  slope = 0.4

Hence, angle to horizontal = arc tan (0.4) = 21.80°
To find the thrust, N, at this point, the arch is cut there and the
equilibrium of the left half is considered (see Fig. 9.7b).

Resolving parallel to N,

$N + (15 \times 10) \cos 68.20 = 375 \cos 21.80 + 300 \cos 68.20$

Hence,

$N = 404$ kN

If we now repeat the same procedure at the support we find the slope is 0.8 and the angle 38.66°. Referring to Fig. 9.7c and resolving parallel to N,

$N = 375 \cos 38.66 + 300 \cos 51.34$

So

$N = 480$ kN

At the crown the arch is horizontal and therefore so is the thrust N. Resolving horizontally will give

$N = 375$ kN

Had we resolved parallel to Q we would have obtained the shear force and taking moments would have yielded M. However, in this instance – a UDL on a parabolic arch – both Q and M are zero everywhere.

A plot of the axial thrust N is given in Fig. 9.7d.

Example 5 (Fig. 9.8a)

Investigate the shear force on either side of the load.
From

$$\text{slope} = \frac{4h}{L^2} (L - 2x)$$

we have

$$\text{Slope} = \frac{4 \times 16}{100 \times 100} (100 - 2 \times 25) = 0.32$$

Hence,

$\theta = 17.74°$

Resolving parallel to Q (Fig. 9.8b), just above the 100 kN load, we obtain

$Q + 78.1 \cos 72.26 + 100 \cos 17.74 = 75 \cos 17.74$

Hence,

$Q -44.0$ kN   (just above the load).

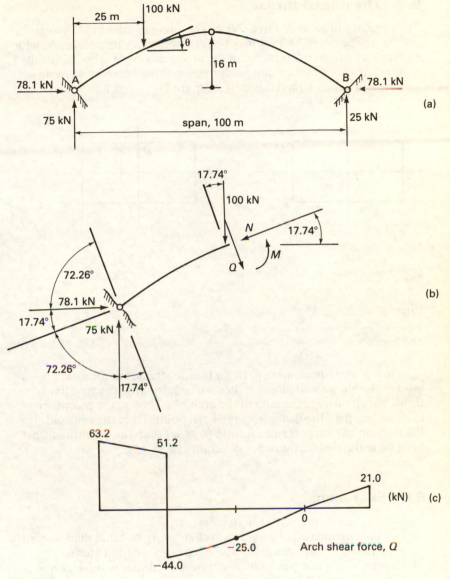

Fig. 9.8

Just below the load the equation will not contain the term 100 cos 17.74 and so we find

$Q = 51.2$ kN   (just below the load).

A plot of the shear force Q is given in Fig. 9.8c.

164

9.7 The line of thrust

The presence of an axial force (N) and bending moment (M) may be seen in a different light; they may be replaced by a single force, N, with eccentricity e from the centroid of the arch section. In other words the moment M may be taken into account be making N eccentric by an amount e. To understand this step compare Figs. 9.9a and 9.9b.

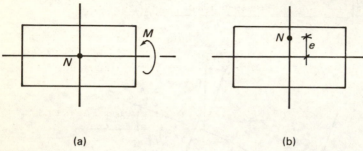

(a)   (b)

Fig. 9.9

Clearly $Ne = M$

and hence, $e = \dfrac{M}{N}$

Now if we require there to be no tensile stresses in the arch rib – a most desirable state of affairs – then we require this eccentric line of thrust to run through the core of the arch rib. That is, the eccentricity must be less than the distance of the kern points from the centroid. If the arch rib has a rectangular section ($b \times d$) then the eccentric thrust must lie in the middle third. So we require, $e < d/6$.

9.8 Exercises

1 Outline briefly the way an arch behaves.
2 In what circumstances would an arch be more suitable than a beam?
3 How many support reactions are there in a three-pin arch?
4 Explain why a two-pin arch cannot be solved simply by resolving forces and taking moments.
5 Describe an arch bridge that you have actually seen and illustrate your description with a sketch.
6 A three-pin parabolic arch spans 80 m and its centre pin has a rise of 18 m above the horizontal line joining the supports. Determine the reactions when seven vertical loads of 100 kN act at 10 m centres across the span. Hence calculate the bending moments under each load.

7 A symmetrical three-pin parabolic arch spans 48 m and has a central rise of 7.8 m. A load of 150 kN acts at the left quarter point. Calculate the bending moment and thrust at the right quarter point.

8 It is required to know how the horizontal thrust (H) at the abutments varies with the rise (r) for a 100 m span symmetrical three-pin arch carrying 10 kN/m. Calculate H in terms of r and hence plot a graph of H against r for $r = 10, 11, \ldots, 20$ m.

9 Using a suitable notation deduce a formula which gives the horizontal thrust for a symmetrical three-pin arch supporting a UDL across the whole span.

10 A three-pin arch has a rise of 10 m and span of 50 m. The arch is symmetrical and parabolic. A load of 200 kN acts 100 m to the left of the centre pin. Calculate the reactions. Hence determine the bending moment (M) and thrust (N) at 5 m centres in the left half of the arch.

11 Analyse the shear and thrust at the left support of a symmetrical three-pin arch when it carries a load at the right quarter point. Take $W = 140$ kN, $L = 90$ m, $r = 15$ m.

12 Determine the reactions of the three-pin arch shown in Fig. 9.10.

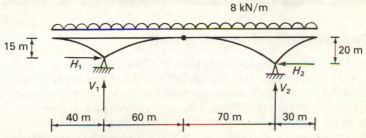

Fig. 9.10

13 Loads of 10 kN each at 5 m centres act on the arch shown in Fig. 9.4. Calculate the reactions and determine the bending moments under each load. Hence sketch a BM diagram for the arch. Assume both halves of the arch are parabolas with their highest points at the crown, C.

14 Show how the shear force, Q, varies in the arch of Fig. 9.6a.

15 The arch of Fig. 9.8 supports a UDL of 12 kN/m on the right half of the arch. Calculate the maximum sagging and hogging moments in the arch.

Chapter 10

Walls

10.1 The special features of a wall

The wall is the structure which – more than any other – is subject to horizontal loads. Certainly it has its own self weight and possibly other vertical loads but its main purpose is to resist loads which are horizontal. These horizontal loads come in three main ways – from wind, from water and from retained earth or other granular material. The horizontal forces from these three – air, water, earth – are considered in the following sections.

10.2 Horizontal forces due to wind

Figure 10.1 shows a wall acted on by a wind of speed v m/sec. The wind will produce a uniform pressure, p, on the side of the wall which in turn may be replaced by a resultant force F acting halfway up the wall. We now give the connection between v, p and F.

Formula 10.1
Pressure due to wind
$p = 0.6v^2$ N/m² where v is the wind speed in m/sec.

Formula 10.2
Force due to wind
$F = ph$ N (acting halfway up) where h is the height of the wall and the length of the wall is 1 metre.

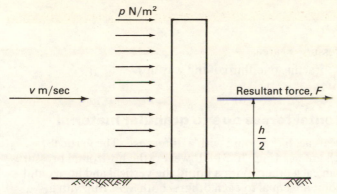

Fig. 10.1

10.3 Horizontal forces due to water

Figure 10.2 shows a wall with water of depth h acting on one side. The pressure on the wall varies from zero at the top to p N/m² at the base and the resultant force to which these pressures are equivalent acts at a height of one-third of the water depth. We now give the equations for the pressure and force on the wall. It is, of course, extremely important to keep the distinction between pressure (N/m²) and force (N) clearly in your mind.

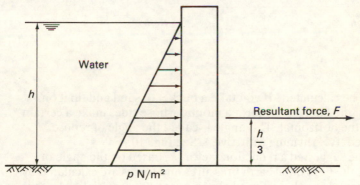

Fig. 10.2

Formula 10.3

Pressure due to water

$$p = \gamma g h \text{ N/m}^2 \qquad (\text{at depth } h)$$

where γ (pronounced 'gamma') is the density of water, 1000 kg/m³, and g is the acceleration due to gravity, 9.81 m/sec².

Corollary The average pressure on the wall is $\frac{1}{2}\gamma g h$.

Formula 10.4
Force due to water

F = average pressure × area

   = $\frac{1}{2}\gamma gh^2$ N   (acting one-third of the way up).

10.4 Horizontal forces due to granular material

Granular materials act like a liquid with a difference. The vertical pressures in both are γgh (Fig. 10.3). However, the horizontal pressures are different from each other. With a liquid the vertical and horizontal pressures at a point are equal to each other – both γgh. But with a granular material the horizontal pressure is less – it is a factor k times the vertical pressure (look again at Fig. 10.3). On what does k depend?

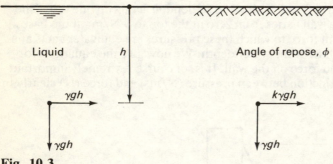

Liquid h Angle of repose, ϕ

γgh $k\gamma gh$

γgh γgh

Fig. 10.3

How can it be calculated? If you take a bucket of sand and tip it on to a spot on the floor, it will form a mound whose sides make a certain angle with the horizontal. This angle is called the angle of repose, ϕ (pronounced 'fye' rhyming with 'dye'). See Fig. 10.4

   Rankine suggested a reduction factor, k, based on the angle of repose. All the horizontal earth pressures and forces are calculated by the same formulae as have already been quoted but multiplied by the reduction factor, k.

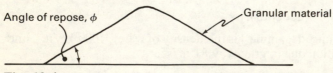

Angle of repose, ϕ Granular material

Fig. 10.4

Formula 10.5

Rankine's factor for granular materials $k = \dfrac{1 - \sin \phi}{1 + \sin \phi}$

Mention must now be made of the so-called 'passive' earth pressures. So far this section has dealt with the earth pressures that are actually acting on a wall – they are called the 'active' earth pressures. So what are the passive pressures? Simply this: the pressures required to push the earth back. So we have:

active pressure is required to hold the earth in position,
passive pressure is required to push the earth back.

Both pressures are horizontal pressures. And, of course, the passive pressure is very much larger than the active pressure (although the names do not suggest this). The same formula is used for calculating passive pressures but with a different constant k.

Formula 10.6

Rankine's factor for passive pressures $k = \dfrac{1 + \sin \phi}{1 + \sin \phi}$

Corollary The values of k set out in Table 10.1 may be found by putting $\phi = 0, 10, \ldots$, degrees into Formulae 10.5 and 10.6.

Table 10.1

ϕ	0	10	20	30	40	50 degrees
k (active)	1.000	0.704	0.490	0.333	0.217	0.132
k (passive)	1.000	1.42	2.04	3.00	4.61	7.58

Formula 10.7

Force due to granular materials

$F = \frac{1}{2}k\gamma gh^2$ N   (acting one-third of the way up)

active or passive – just use the right value of k!

Example 1
A wall 2.4 m high is subjected to a wind speed of 15 m/sec. Calculate the pressure on the wall and the force on a 1 m length of wall.

From
$$p = 0.6v^2$$
we have the wind pressure, $p = 0.6 \times 15^2 = 135$ N/m^2.

And from

$$F = ph$$

we have the wind force, $F = 135 \times 2.4 = 324$ N (on 1 m length of wall)

Example 2

The vertical face of a dam rises 30 m above the river bed. Determine the pressure, p, at the base of the dam and the resultant water force, F, on the dam when the depth of water is $h = 5, 10, 15, 20, 25, 30$ metres.

We know

$$p = \gamma g h$$

and

$$F = \tfrac{1}{2}\gamma g h^2 \qquad \text{(acting } h/3 \text{ from the base).}$$

Now when something has to be worked out a number of times we must polish up the formula as much as possible before we substitute for the variables.

So, putting $\gamma = 1000$ kg/m^3 as the density of water

and $\quad g = 9.81$ m/sec^2

We now obtain,

$$p = 9810h \text{ N/m}^2 = 9.81h \text{ kN/m}^2$$

and,

$$F = 4905\, h^2 \text{ N} = 4.905\, h^2 \text{ kN}$$

Substituting now for h we may obtain Table 10.2.

Table 10.2

h (metres)	5	10	15	20	25	30
p (kN/m^2)	49	98	147	196	245	294
F (kN)	123	491	1104	1962	3066	4415

Example 3

A retaining wall supports some granular material of density 1600 kg/m^3 to a depth of 4.2 m. The angle of repose of the material is 32°. Calculate the resultant horizontal force on the wall.

From

$$F = \tfrac{1}{2}k\gamma g h^2$$

and

$$k = \frac{1 - \sin\phi}{1 + \sin\phi}$$

we have

$$k = \frac{1 - \sin 32}{1 + \sin 32} = 0.307$$

and hence,

$$F = 0.5 \times 0.307 \times 1600 \times 9.81 \times 4.2^2$$
$$= 42\ 500\ \text{N}$$

or, $F = 42.5$ kN.

10.5 Surcharge

The term 'surcharge' refers to an additional pressure on the top surface of the soil (or other material). The effect of surcharge is to introduced a second force on the wall (Fig. 10.5).

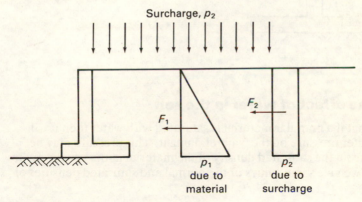

Fig. 10.5

Forces due to material with surcharge

$F_1 = \frac{1}{2}\gamma g h^2$ (acting one-third of the way up)

$F_2 = p_2 h$ (acting one-half of the way up) where p_2 is the surcharge pressure.

Of course, these two forces may be combined to give a resultant force acting somewhere between one-third and one-half of the way up the wall.

10.6 Line loads

The top surface of the retained material may carry a line load of W kN/m which may be taken to exert a horizontal force on the wall (Fig. 10.6).

Force due to a line load

$$F = kW \qquad \text{where} \qquad k = \frac{1 - \sin \phi}{1 + \sin \phi}$$

The point at which this force acts may be taken as the point in which a line at 40° to the horizontal coming from the load strikes the wall. These results are of course only approximate and no proof is given here.

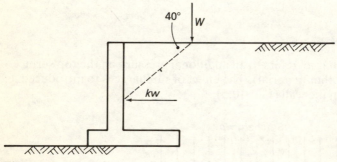

Fig. 10.6

10.7 The effect of water in the soil

It is clear that if a granular material is saturated with water then it will exert a greater pressure on the wall retaining it. This pressure may be calculated from the saturated density of the material in the usual way. In Table 10.3 we give some values of the normal and saturated densities of

Table 10.3

Material	Density kg/m^3	Saturated Density kg/m^3
Gravel	1800	2100
Coarse sand	1900	2100
Fine sand	2000	2100
Limestone	1600	2000
Chalk	1100	1500
Ash	800	1400

some common granular materials. The figures are only typical and the actual values may be found to lie up to 15 per cent either side of these figures.

10.8 The pressures under a wall

The forces on a wall – its self weight and any horizontal forces from wind, water or soil – will be transmitted through the wall to the ground. It is then of great importance to find the resulting ground pressures and compare them with the permissible ground pressures. Let us see how this is done. We have met the principles before in the chapter on stress (Chapter 5).

How to find base pressures under a wall
1 Calculate the total vertical load, W.
2 Calculate the moment of all the forces about the centre of the base, M.

Then, base pressures, $$p = \frac{W}{A} \pm \frac{M}{Z}$$

where,

$A = bd$ is the area of the base

$Z = \dfrac{bd^2}{6}$ is the section modulus of the base

and $b = 1$ m length of wall

d = the width of the base of the wall.

Example 4 (Fig. 10.7)
Determine the base pressures under the given wall. Assume density of wall to be 2400 kg/m^3.

Assuming a 1 m length of wall,

wind force, $F = 100 \times 3 \times 1/10^3 = 0.3$ kN acting 1.5 m up.

Weight of wall, $W = 2400 \times 9.81 \times 3 \times 0.225 \times 1/10^3 = 16.9$ kN

Moment about base,

$M = 0.3 \times 1.5 = 0.45$ kN m

And for the base area we have

area, $A = 1 \times 0.225 = 0.225$ m^2

and modulus, $Z = \dfrac{bd^2}{6} = \dfrac{1 \times 0.225^2}{6} = 8.43 \times 10^{-3}$ m^3

174

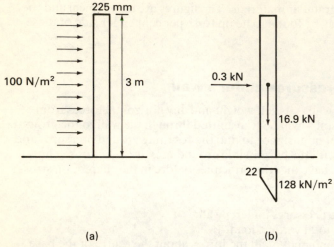

225 mm

100 N/m²

3 m

0.3 kN

16.9 kN

22

128 kN/m²

(a)

(b)

Fig. 10.7

Hence the base pressures,

$$p = \frac{W}{A} \pm \frac{M}{Z}$$

So, $p = 75.0 \pm 53.3 = 128$, and 22 kN/m²   (see Fig. 10.7b).

Example 5 (Fig. 10.8)
Determine the base pressures under this concrete gravity dam when the reservoir is full and when it is empty. The density of concrete is 2400 kg/m³.

The water force, $F = \frac{1}{2}\gamma g h^2$

$$= 0.5 \times 1000 \times 9.81 \times 6^2/10^3$$

$$= 177 \text{ kN acting 2 m up.}$$

Weight of dam, $W = 2400 \times 9.81 (6 \times 1 \times 5/2)/10^3 = 353$ kN.

The centre of gravity of the dam is $\bar{x}$ from the water face, where

$$\bar{x} = \frac{\Sigma Ax}{\Sigma A}$$

$$= \frac{6 \times 0.5 + 9 \times 2}{15}$$

$$= 1.4 \text{ m}$$

Hence the weight is eccentric $e = 0.6$ m from the centre of the base (Fig. 10.8b).

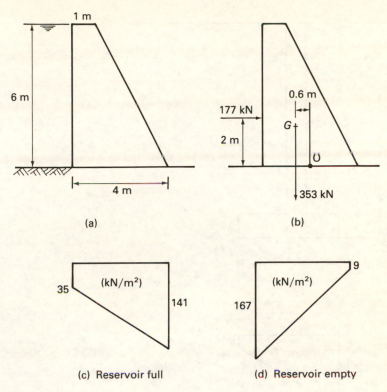

Fig. 10.8

The moment of all the forces about the centre of the base (O) is,

$M = 177 \times 2.0 - 353 \times 0.6$

$\quad = 142$ kN m   clockwise.

Putting $A = 4.0$ m², $Z = 2.67$ m³ gives the base pressures,

$$p = \frac{W}{A} \pm \frac{M}{Z} = 88 \pm 53 = 141, \text{ and } 35 \text{ kN/m}^2.$$

If the reservoir is empty,

$\quad\quad M = 353 \times 0.6 = 212$ kN m anticlockwise.

So, $p = \dfrac{W}{A} \pm \dfrac{M}{Z} = 88 \pm 79 = 167, \text{ and } 9 \text{ kN/m}^2$

These results are shown in Figs. 10.8c, d.

Example 6 (Fig. 10.9)

Determine the pressure distribution under the retaining wall shown. For the retained material take $\gamma = 2000$ kg/m³, $\phi = 35°$.

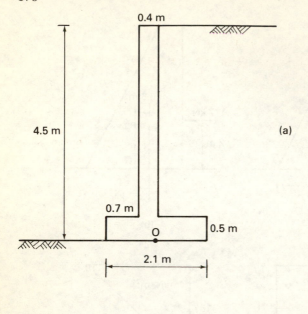

(a)

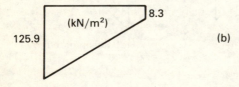

(b)

Fig. 10.9

Let us first calculate the weights of the various elements – the stem of the wall, the base of the wall and the material vertically above the base.

Element **Weight**
Stem $4.0 \times 0.4 \times 1.0 \times 2400 \times 9.81/10^3$ =   37.7 kN
Base $2.1 \times 0.5 \times 1.0 \times 2400 \times 9.81/10^3$ =   24.7 kN
Material $4.0 \times 1.0 \times 1.0 \times 2000 \times 9.81/10^3$ =   78.5 kN
                                     Total, $W = 140.9$ kN

Now the horizontal earth force is

$$F = \tfrac{1}{2}k\gamma gh^2$$

with $k = \dfrac{1 - \sin \phi}{1 + \sin \phi} = 0.271$

So,

$F = 0.5 \times 0.271 \times 2000 \times 9.81 \times 4.5^2 \times 10^{-3}$

$= 53.8$ kN acting 1.5 m above the lower edge of the base.

We need now to take moments about the centre of the base:

$M = 53.8 \times 1.5 + 37.7 \times 0.15 - 78.5 \times 0.55$

$= 43.2$ kN m anticlockwise.

Putting, $W = 140.9$ kN, $M = 43.2$ kN m, $A = 2.1$ m^2, $Z = 0.735$ m^3 into

$$p = \frac{W}{A} \pm \frac{M}{Z}$$

gives the base pressures, $p = 67.1 \pm 58.8 = 125.9$, and 8.3 kN/m^2 as shown in Fig. 10.9b.

10.9   Uplift and how to avoid it

If the horizontal forces on a wall are gradually increased there will come a point when one edge of the wall will begin to lift. The wall is not about to overturn but the compressive forces along the edge are reduced to zero and – since there is no tensile strength between the wall and the ground – the edge will lift up. It is an undesirable state of affairs but before looking at the way to avoid it let us see how to recognise it when it occurs.

Uplift begins when $\dfrac{M}{Z} = \dfrac{W}{A}$

Putting

$e = \dfrac{M}{W}$ and $Z = \dfrac{bd^2}{6}$, $A = bd$

gives,

$e = \dfrac{d}{6}.$

How to recognise uplift?

1 Calculate $e = \dfrac{M}{W}$.

2 If $e > \dfrac{d}{6}$ there is uplift.

178

Look at Fig. 10.10a. The forces on the wall may be reduced to the horizontal load (F) and the vertical load (W) eccentric e from the centre of the base. The resultant of these two forces will cut the base a distance e from its centre. What pressure diagram will be equivalent to this? The pressure diagram, which only resists the vertical loads, has simply to be equivalent to W eccentric e. For the case of uplift, the diagram must therefore be a triangle of area W with its centroid eccentric e from the base centre (see Fig. 10.10b).

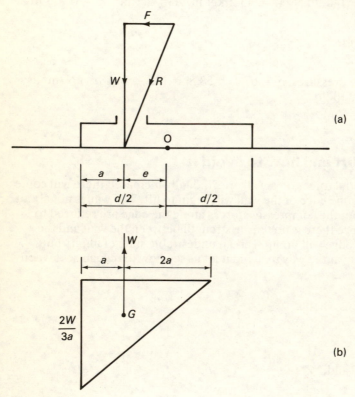

Fig. 10.10

How to draw the uplift pressure diagram

Base length $= 3a$

Toe pressure $= \dfrac{2W}{3a}$

where $a = \dfrac{d}{2} - e$

Example 7 (Fig. 10.7)

Investigate the pressure distribution under the wall if the wind pressure is 150 N/m². At what wind pressure does uplift begin?

With a wind pressure of 150 N/m²,

wind force, $\quad F = 150 \times 3 \times 1/10^3 = 0.45$ kN

Hence,

$M = 0.45 \times 1.5 = 0.675$ kN m

But,

$W = 16.9$ kN

Hence,

$$e = \frac{M}{W} = 0.0400 \text{ m} = 40.0 \text{ mm}$$

But,

$$\frac{d}{6} = \frac{225}{6} = 37.5 \text{ mm}$$

So,

$$e > \frac{d}{6} \quad \text{and we have uplift.}$$

Hence,

$$a = \frac{d}{2} - e = 112.5 - 40.0 = 72.5 \text{ mm}$$

So for the uplift pressure diagram we obtain,

base length $= 3 \times 72.5 = 218$ mm

and

$$\text{toe pressure} = \frac{2W}{3a} = \frac{2 \times 16.9}{3 \times 0.0725} = 155 \text{ kN/m}^2$$

Uplift will just begin when $\quad e = \dfrac{M}{W} = 37.5$ mm

So,

$M = 0.0375 \times 16.9 = 0.634$ kN m

Working backwards,

wind force, $F = 0.634/1.5 = 0.426$ kN

180

Hence,

critical wind pressure, $p = 426/(3 \times 1) = 141 \text{ N/m}^2$

in which situation the initial toe pressure is twice the dead load pressure of W/A – in this case 150 kN/m^2.

To avoid uplift the base of the wall needs to be extended (so increasing both A and Z). In the case of a retaining wall the heel should be extended further under the retained material and the calculations started again with the new dimensions of the wall.

10.10 The factor of safety against overturning

A wall needs to be some way from overturning for it to be safe. How can its safety be measured?

Factor of safety against overturning $\qquad \lambda_t = \dfrac{\text{restoring moment}}{\text{overturning moment}}$

which must be greater than 2 for safety.

Note: Moments must be taken about the point about which it will turn.

Example 8 (Fig. 10.7)
Calculate the factor of safety against overturning.

Since the wall will turn about its right-hand lower edge, this is the point about which to take moments.

Restoring moment = $16.9 \times 0.1125 = 1.89$ kN m
Overturning moment = $0.3 \times 1.5 = 0.45$ kN m

Hence,

factor of safety, $\qquad \lambda_t = \dfrac{1.89}{0.45} = 4.2,$ $\qquad$ which is satisfactory.

Example 9 (Fig. 10.8b)
Calculate the factor of safety against overturning.

Taking moments about the downstream edge,
restoring moment = $353 \times 2.6 = 918$ kN m
overturning moment = $177 \times 2 = 354$ kN m

Hence,

$\lambda_t = 2.6$, which is satisfactory.

Example 10 (Fig. 10.11)

Calculate the factor of safety against overturning.

Clearly,

$$\lambda_t = \frac{148.7 \times 1.29}{53.8 \times 1.4} = 2.5$$

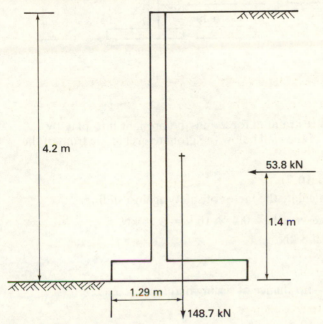

Fig. 10.11

10.11 The factor of safety against sliding

The same principle applies in calculating this factor of safety.

Factor of safety against sliding

$$\lambda_s = \frac{\text{restraining forces}}{\text{sliding forces}}$$

which must be greater than 2 for comfort.

Note 1: The main restraining force comes from the frictional resistance under the base of the wall (Figure 10.12a).

Frictional force $= \mu W$

where μ is the coefficient of friction (0.4–0.6 is common), and W is the total weight.

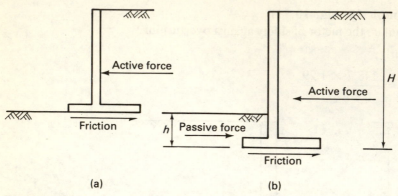

Fig. 10.12

Note 2: A second restraining force may be brought into play by lowering the base of the wall below the ground level at the front of the wall (Fig. 10.12b).

Example 11 (Fig. 10.7)

Taking $\mu = 0.4$ calculate the factor of safety against sliding.

Restraining force = $\mu W = 0.4 \times 16.9 = 6.76$ kN
Sliding force = 0.3 kN

Hence,

$$\lambda_s = \frac{6.76}{0.3} = 23 - \text{no chance of sliding here!}$$

Example 12 (Fig. 10.8)

With $\mu = 0.5$ calculate λ_s.

Restraining force = $0.5 \times 353 = 177$ kN
Sliding force = 177 kN

Hence,

$$\lambda_s = \frac{177}{177} = 1.00 - \text{which means the structure is on the point of sliding}$$

Clearly the structure is just not big enough. By lowering the base the frictional resistance is increased and the passive resistance of the soil would be brought into play at the downstream face.

Example 13 (Fig. 10.11)

Calculate λ_s with $\mu = 0.4$.

$$\text{Simply, } \lambda_s = \frac{0.4 \times 148.7}{53.8} = 1.11 - \text{no good.}$$

The situation would be improved in this case by lowering the base to increase the resistance to sliding and lift λ above 2 as in the next example.

Example 14 (Fig. 10.12b)

Investigate the factor of safety λ_s as the base is lowered. Use same data as Example 6 (Fig. 10.9a).

Using $\phi = 35°$, we have

$$k \text{ (active)} = \frac{1 - \sin \phi}{1 + \sin \phi} = 0.271$$

$$k \text{ (passive)} = 1/0.271 = 3.69$$

Hence,

active force, $\quad F_a = \frac{1}{2}k\gamma gH^2 = 2.66\ H^2$ kN

passive force, $\quad F_p = \frac{1}{2}k\gamma gh^2 = 36.2\ h^2$ kN

and

friction force, $F = 0.4W$

where W = total weight of earth and wall above the base.

Finally,

factor of safety against sliding, $\lambda_s = \dfrac{F_p + 0.4W}{F_a}$

Taking some values of h until we reach $\lambda_s = 2.0$ or more we obtain the results shown in Table 10.4.

Table 10.4

Depth of base h	Passive Force	Frictional Force	Active Force	Factor of Safety λ_s
0.0	0.0	54.4	53.8	1.01
1.0	36.2	67.7	71.8	1.45
1.5	81.4	74.4	86.3	1.81
1.7	104.5	77.1	92.4	1.97
1.8	117.2	78.4	95.6	2.05
metres	*kN*	*kN*	*kN*	

Notice how rapidly the passive force rises compared with the active force. The frictional force is increasing at the same time but only slowly. With a little trial and error we can soon arrive at $h = 1.8$ m with $\lambda_s = 2.05$, which is satisfactory.

184

10.12 Exercises

Take, concrete density 2400 kg/m³, water density 1000 kg/m³.

1. Name one special feature of a wall.
2. State three quantities to be calculated in the analysis of a wall.
3. What pressure does a wind of speed v m/sec exert?
4. Quote the formula for the pressure at a depth h in a liquid of density γ.
5. A granular material of height h is retained by a vertical wall. Quote the formula for the horizontal force on the wall defining any terms you use.
6. Define the angle of repose, ϕ.
7. What is Rankine's factor for active pressures?
8. What is Rankine's factor for passive pressures?
9. Distinguish between active and passive pressures.
10. Explain the term 'surcharge'.
11. What effect does water have in the soil?
12. Quote the formula for the pressures under a wall if the weight of the wall is W and the moment of all the forces about the centre of the base is M.
13. How can you tell if there will be uplift under a wall?
14. Quote the formula for the pressure under the toe of a wall when there is uplift. What is the length of the pressure diagram under the wall?
15. Define the two factors of safety connected with walls. What is the minimum value that they should be?
16. Calculate the pressure under a concrete wall 300 mm thick and 2 m high.

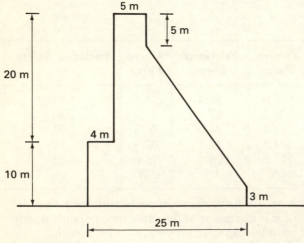

Fig. 10.13

17 A wall is 225 mm thick and of height 2.5 m. A wind of speed 12 m/sec exerts a pressure on one side. Calculate the base pressures and determine the factor of safety against overturning, λ_t.

18 A mass concrete wall 1 m thick and 3 m high has water rising up one side of it. Determine the way the base pressures change taking water heights from zero increasing at 0.5 m intervals. At what height will uplift occur? At what height will the wall topple?

19 A section through Kainji Dam in Nigeria is shown in Fig. 10.13. Determine the weight of 1 m length of a dam and locate its centre of gravity. If the water depth is h, calculate the base pressures for $h = 0, 10, 20, 30$ m. Allowing a coefficient of friction of 0.4, determine the factor of safety against sliding, λ_s, for these values of h. Hence plot a graph of λ_s against h and comment on it. How does λ_t against h compare with λ_s against h?

20 A retaining wall (Fig. 10.14) supports material of density $\gamma = 1600$ kg/m³, angle of repose, $\phi = 40°$ and coefficient of friction, $\mu = 0.5$. Calculate the base pressures and the two safety factors λ_t and λ_s. Should the base be lowered to raise λ_s above 2? Show how λ_s varies as the base is shortened and draw a graph of this variation.

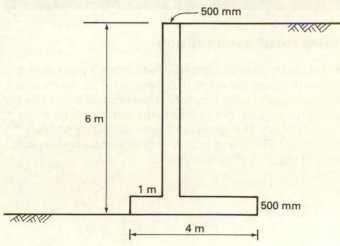

Fig. 10.14

Chapter 11

Foundations

11.1 The purpose of a foundation

It has been pointed out in Chapter 1 that the foundation's purpose is to transfer the loads in the structure safely to the ground. The significance of this is often masked by the units we use for the permissible stresses in the structure and in the ground. For example, the working stress in concrete is about 8 N/mm². The permissible ground bearing pressure may be, say, 200 kN/m². But how do these stresses compare with each other? Which is the larger? And by how much?

We may show

$$1 \text{ N/mm}^2 = 1000 \text{ kN/m}^2$$

So

allowable concrete stress $= 8000 \text{ kN/m}^2$

But,

allowable ground pressure $= 200 \text{ kN/m}^2$

There is 40 times too great a stress in the column for it to rest directly on the ground!

And, incidentally, the same argument applies to a steel column resting on a concrete base.

Converting again to kN/m² for convenience,

allowable steel stress   = 155 000 kN/m²
allowable concrete stress =   8 000 kN/m²
allowable ground stress   =   200 kN/m².

So there is 20 times too great a stress in steel for it to rest on concrete. These figures show why it is necessary for the structure to spread out and increase in area as it comes to rest on the ground.

Example 1

What area of steel, concrete and ground is required to support 1000 kN using the above stresses?

From

$$f = \frac{W}{A}$$

we have

$$A = \frac{W}{f}$$

So,

for steel: $A = \dfrac{1000}{155\,000} \times 10^6 = 6500 \text{ mm}^2$

for concrete: $A = \dfrac{1000}{8000} \times 10^6 = 125\,000 \text{ mm}^2$

for ground: $A = \dfrac{1000}{200} \times 10^6 = 5\,000\,000 \text{ mm}^2$

These gives squares of sides, 80 mm, 350 mm, 2240 mm approximately. Draw these squares to scale one inside the other.

11.2   The usual base pressures

The pressures under a foundation may be found just as they are found under a wall.

How to find the pressures under a foundation
1 Find the total load, W.
2 Find the moment of all the forces about the centre of the base, M.

Then,

base pressures, $p = \dfrac{W}{A} \pm \dfrac{M}{Z}$

188

where, A is the area of the base,

and, $\qquad Z = \dfrac{bd^2}{6}$

is the section modulus of the base and (note this) b is parallel to the axis about which M acts.

Example 2

A column carries an axial load of 1200 kN. Calculate the size of base to take this load if the permissible ground pressure is 220 kN/m².

From

$$f = \frac{W}{A} \qquad \text{we have} \quad A = \frac{W}{f}$$

So,

$\qquad$ required area, $\qquad A = \dfrac{1200}{220} = 5.45 \text{ m}^2$

So a base 2.4 m by 2.4 m is satisfactory.

Example 3 (Fig. 11.1)

Calculate the pressure distribution under the base.

Clearly,

$A = 2.25 \times 2.25 = 5.06 \text{ m}^2$

$Z = \dfrac{2.25 \times 2.25^2}{6} = 1.90 \text{ m}^3$

Hence,

$\qquad$ base pressures, $\qquad p = \dfrac{W}{A} \pm \dfrac{M}{Z}$

$$= \frac{870}{5.06} \pm \frac{160}{1.90}$$

$$= 256, \text{ and } 88 \text{ kN/m}^2$$

which are shown in Fig. 11.1b.

Example 4 (Fig. 11.2)

A square foundation 2 m × 2 m carries a load through its centroid of 600 kN. In addition, moments of 80 kN m and 60 kN m act about the OX and OY axes. Determine the stresses at the four corners.

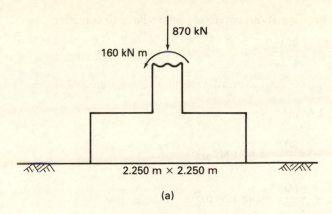

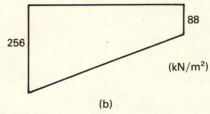

Fig. 11.1

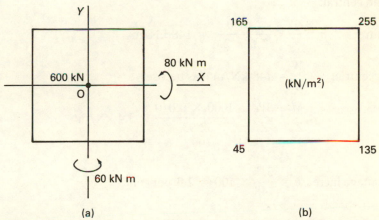

Fig. 11.2

190

Here we have bending about two axes. So we have to consider

$$p = \frac{W}{A} \pm \left(\frac{M}{Z}\right)_x \pm \left(\frac{M}{Z}\right)_y$$

So,

$$\frac{W}{A} = \frac{600}{4} = 150 \text{ kN/m}^2$$

About OX, $\quad \dfrac{M}{Z} = \dfrac{80}{1.333} = 60 \text{ kN/m}^2$

About OY, $\quad \dfrac{M}{Z} = \dfrac{60}{1.333} = 45 \text{ kN/m}^2$

Hence,

$$p = 150 \pm 60 \pm 45 \text{ kN/m}^2$$

Taking each possibility in turn,

$$p = 255, 165, 135, 45 \text{ kN/m}^2$$

as shown in Fig. 11.2b.

Example 5
A load of 1000 kN which should have been centrally placed is eccentric 10 mm along one axis of a 2.6 m × 2.6 m square foundation. By what percentage is the ground pressure increased?

Load central:

pressure, $\quad p = \dfrac{W}{A} = \dfrac{1000}{2.6 \times 2.6} = 148 \text{ kN/m}^2$

Load eccentric: $\quad \dfrac{W}{A} = 148 \text{ kN/m}^2$ (as before)

$$\frac{M}{Z} = \frac{We}{Z} = \frac{1000 \times 0.010}{2.93} = 3.4 \text{ kN/m}^2$$

So

percentage increase $= \dfrac{3.4}{148} \times 100 = 2.3$ per cent.

11.3 Uplift and how to avoid it

Uplift has already been encountered in the previous chapter (section 10.9) and exactly the same principles apply here which we now repeat.

Uplift will begin when $M/Z = W/A$ which for a rectangular section reduces to $M/W = d/6$.

How to recognise uplift?

1 Calculate $e = \dfrac{M}{W}$

2 If $e > \dfrac{d}{6}$ there is uplift.

Note: If the section is not rectangular, the second condition above must simply be changed to say that uplift will occur if the point distant e from the centroid lies beyond the appropriate kern point (see section 4.7). Of course, all we are saying is that if the resultant force passes outside the core then there will be uplift.

How to draw the uplift pressure diagram?

base length $= 3a$

Max. pressure $= \dfrac{2W}{3a}$

where $a = \dfrac{d}{2} - e$ (Fig. 10.10a, b)

It is desirable to avoid uplift whenever possible and this may be done by making the foundation larger.

11.4   The strip foundation

If a number of columns lie in a single line it may be convenient to support them on a single strip foundation (Fig. 1.19) rather than a number of isolated column bases. All that has been said before applies here also: $p = W/A \pm M/Z$ will give the pressure distribution in the usual case.

However, in the same way that it is desirable to avoid uplift, it is also good to avoid any pressure variation under the foundation and to obtain, where possible, a uniform pressure distribution. This of course requires $M = 0$, which is another way of saying that the centre of gravity of the loads must be directly above the centroid of the base of the foundation.

Example 6 (Fig. 11.3a)

The figure shows four column loads supported by a strip foundation. The permissible ground pressure is 200 kN/m^2. If the width of the strip is 1.40 m, determine its length and how it should be positioned so that a uniform pressure distribution occurs under it.

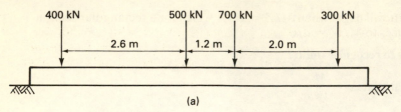

(a)

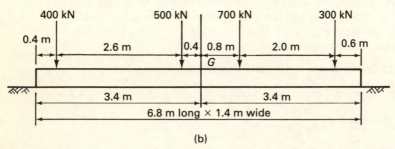

(b)

Fig. 11.3

The centroid of the loads is $\bar{x}$ from the left-hand load, where

$$\bar{x} = \frac{\Sigma Wx}{\Sigma W}$$

$$= \frac{500 \times 2.6 + 700 \times 3.8 + 300 \times 5.8}{400 + 500 + 700 + 300}$$

$$= 3.0 \text{ m}$$

hence the midpoint of the strip must lie 3.0 m to the right of the 400 kN load. From

$$f = \frac{W}{A}$$

we have

$$200 = \frac{1900}{1.4L}$$

So the length of the foundation is

$$L = 6.80 \text{ m}$$

and it should be positioned as shown in Fig. 11.3b.

Example 7 (Fig. 11.4)

The figure shows a set of columns (and their loads) with two movable crane wheels which may act anywhere above the columns. Determine the pressure distribution as the wheels move from one end to the other.

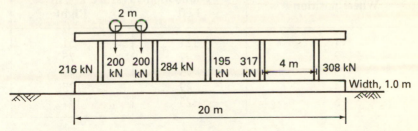

Fig. 11.4

Let us simply take the five cases of the wheels placed symmetrically above each column in turn. The wheels clearly represent 400 kN acting halfway between them.

Total load, $W = 1720$ kN

Base area, $A = 20$ m^2

Hence,

$$\frac{W}{A} = 86.0 \text{ kN/m}^2$$

With the wheels above the left-hand column the moment of all the forces about the centre of the base is,

$$M = 400 \times 8 + 216 \times 8 + 284 \times 4 + 195 \times 0 - 317 \times 4 - 308 \times 8$$
$$= 2332 \text{ kN m}$$

Corresponding values of M when the wheels are above the other columns are,

$$M = 732, -868, -2468, -4068 \text{ kN m.}$$

Putting

$$Z = \frac{bd^2}{6} = \frac{1 \times 20^2}{6} = 66.67 \text{ m}^3$$

gives,

$$\frac{M}{Z} = 35.0, 11.0, -13.0, -37.0, -61.0 \text{ kN/m}^2$$

for the five cases.

Taking care with the signs (let them do the work for you) gives the pressures at the left and right of the foundation strip using $p = W/A \pm M/Z$. The results are given in Table 11.1.

Table 11.1.

Wheel position	Foundation Pressure kN/m²	
	Left	Right
1	121	51
2	97	75
3	73	99
4	49	123
5	25	147

Example 8 (Fig. 11.5)

The Figure shows two columns A and B. A wall is built so close to column B that the foundation may not extend beyond it. Find a suitable base of the trapezoidal shape shown. Allow a permissible ground pressure of 180 kN/m².

The aim is to calculate the distances a and b so that the pressure under the foundation is of uniform value of 180 kN/m². This requires the centre of gravity of the loads to coincide with the centroid of the base area.

The centroid of the trapezium is $\bar{x}$ from side a where,

$$\bar{x} = \frac{a + 2b}{3a + 3b} \cdot h \quad \text{(see Fig. 11.5b)}$$

But the centre of gravity of the loads is $\bar{x}$ from side a also, so,

$$\bar{x} = \frac{\Sigma Wx}{\Sigma W} = \frac{800 \times 0.5 + 1200 \times 4.5}{2000}$$

$$= 2.900 \text{ m}$$

Putting $h = 5.0$ m and rearranging gives,

$$2.9(3a + 3b) = 5.0(a + 2b)$$

or, $3.7a - 1.3b = 0$

so, $b = 2.846a$

Using this and the fact that the pressure under the base is 180 kN/m² we obtain,

$$180 \times \tfrac{1}{2}(a + 2.846a) \times 5.0 = 2000$$

So, $1731a = 2000$

Hence,

$a = 1.160$ m and $b = 3.290$ m.

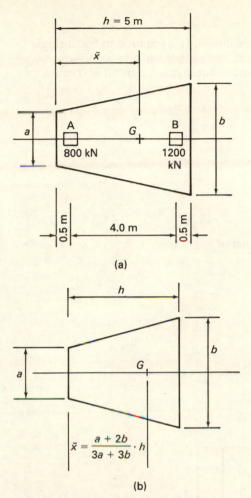

Fig. 11.5

11.5 The raft foundation

On poor soil or soil of varying strength it is often preferable to combine the separate foundations required under each column into one single combined foundation – often called a raft foundation.

The rigidity of such a foundation does not concern us here – although it is very important. We are concerned simply with the base pressures and, as with the strip foundation, how to position the foundation so that there is a uniform pressure under it. Needless to say no new principles are involved and $p = W/A \pm M/Z$ will solve most problems.

Example 9 (Fig. 11.6)

Four columns are spaced apart as shown. It is required to find a single raft foundation which will support these loads and give a uniform pressure on to ground whose permissible stress is only 50 kN/m².

Using $\bar{x} = \dfrac{\Sigma Wx}{\Sigma W}$ and $\bar{y} = \dfrac{\Sigma Wy}{\Sigma W}$

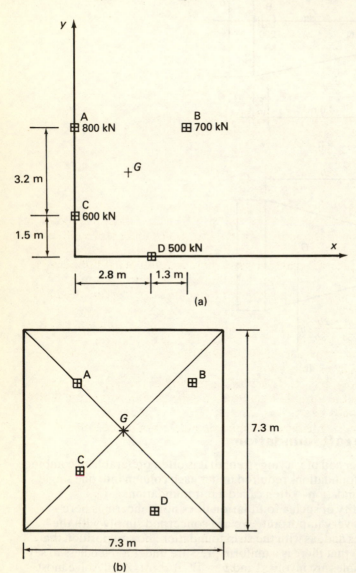

(a)

(b)

Fig. 11.6

we obtain

$$\bar{x} = \frac{500 \times 2.8 + 700 \times 4.1}{2600} = 1.64 \text{ m}$$

$$\bar{y} = \frac{600 \times 1.5 + 800 \times 4.7 + 700 \times 4.7}{2600} = 3.06 \text{ m}$$

The size of the base is found from $f = \dfrac{W}{A}$

which gives,

$$A = \frac{2600}{50} = 52.0 \text{ m}^2$$

Hence, we may chose a raft 7.300 m × 7.300 m.
A dimensioned sketch is given in Fig. 11.6b.

11.6 Exercises

1 Explain briefly why a foundation is necessary.
2 Quote the usual equation for the base pressures under a foundation.
3 Explain how you would check whether or not there was uplift under a foundation.
4 Draw the pressure diagram when there is uplift. What is the length of the base of the diagram? What is the maximum pressure?
5 Explain briefly the terms isolated foundation, strip foundation and raft foundation.
6 Why are pressure differences undesirable under a foundation? How can they be avoided?
7 Quote the formula for the position of the centre of gravity of a set of loads lying in a straight line.
8 A set of loads $W_1, \ldots, W_n$ have coordinates $(x_1, y_1), \ldots, (x_n, y_n)$. Quote the formula which will give the coordinates of the centre of gravity, $G(\bar{x}, \bar{y})$.
9 Determine the pressure distribution under each of the isolated column bases shown in Fig. 11.7.

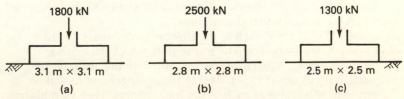

1800 kN   2500 kN   1300 kN

3.1 m × 3.1 m   2.8 m × 2.8 m   2.5 m × 2.5 m

(a)   (b)   (c)

Fig. 11.7

10 Calculate suitable square foundations to support loads of 1200 kN, 1600 kN, 2000 kN on ground whose permissible stresses are respectively 200 kN/m², 160 kN/m², 120 kN/m².

11 Construct a chart with:

x axis: column loads (W), $0 - 4000$ kN

y axis: ground pressures (p), $0 - 400$ kN/m²

and the body of the chart with lines for suitable square foundations d metres by d metres.

Hints: Since $p = W/d^2$

take $d = 1.0, 2.0, 3.0, 4.0$ and 5.0 metres

and plot the resulting straight lines.

12 Three foundations 2.2 m × 2.2 m, 1.6 m × 14.0 m and a trapezoidal base with parallel sides 2.8 m, 4.2 m distant 6.0 m apart each rest on ground whose permissible bearing pressure is 225 kN/m². Calculate the load that each foundation may take at its centroid.

13 Determine the pressure distribution under the bases shown in Fig. 11.8.

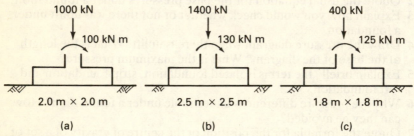

Fig. 11.8

14 A load of 1500 kN is supported by a square base 2.6 m × 2.6 m. What is the maximum moment it can withstand if there is to be no uplift?

15 Calculate the length of a suitable strip foundation of width 1.5 m to support two columns (1230 kN, 1370 kN), 4 m apart. Allow ground pressure of 250 kN/m². Draw a dimensioned sketch to show how the foundation is placed under the column.

16 Loads of 4000 kN act at each of the three coordinates A (4, 10), B (10, 12), C (8, 4) metres. The foundation under the loads is proposed as a single square raft foundation with the ends of one diagonal at (2, 2) and (14, 14). Sketch the arrangement to scale. Calculate the pressures at each of the four corners. How should the foundation be placed to give a uniform pressure underneath?

17 Calculate the dimensions of a trapezoidal base to support two columns (1570 kN, 2110 kN), 3.6 m apart. The base may extend along the line of the columns no further than 0.3 m beyond the centre of either column. Allow a ground pressure of 320 kN/m².

18 Three columns (1000 kN, 800 kN, 700 kN) lie in a straight line with 3 m and 2 m between them. A strip foundation is proposed of width 1.2 m. Determine its length and position so as to give a uniform pressure of 280 kN/m² under it. If the left-hand load increases by 10 per cent determine the new pressure distribution.

19 A building of total weight 44 000 kN is to be supported on 16 columns arranged in a square grid at 6 m centres in both directions. Determine the foundation area required for each of the three permissible ground pressures of $p = 50$ kN/m², 150 kN/m² and 250 kN/m². Draw to scale the layout of the columns (they may be taken to be 400 mm × 400 mm) and show how the required foundation area may be obtained in each of the three cases.

20 A circular chimney 55.0 m high rests on a square foundation 8 m by 8 m by 2 m thick. The chimney section has an internal diameter of 4.0 m and an external diameter of 5.0 m. Allowing a density of 2500 kg/m³ for the chimney, determine (*a*) the stress in the chimney wall at its base, and (*b*) the pressure under the foundation. You may assume the foundation weighs 3200 kN. If the wind now blows with a speed of 15 m/sec on the side of the chimney determine the new pressures under the foundation. (Assume $p = 0.6\ v^2$ N/m² for the wind pressure and that it acts upon the chimney rectangle of 5 m × 55 m).

Part III

Concrete

Concrete

Chapter 12

Reinforced concrete

12.1 Elastic analysis and ultimate load analysis

There are two important and quite distinct loading cases associated with a structure. The first is the normal everyday loading that the structure is designed to support. This is called the working load and includes the live load as well as dead load. Furthermore, any deflections that are produced by these loads will disappear if the loads are removed. The stresses in the structure are then said to be in the *elastic* range. The term *elastic* simply means that the deflections will vanish if the stresses are removed. The analysis of the structure in such a state is called elastic analysis. Such is the first type of loading and its analysis.

The second loading is quite a different matter. It is the load which will cause the structure to collapse. Removal of the loads in such a case will in no way result in the structure returning to its original undeflected form. Such loading is called the ultimate load of the structure and the analysis of the collapse of the structure or the ultimate strength of one of its members is called ultimate load analysis.

Let us summarise this:

Two types of analysis
1 Elastic analysis of working load behaviour.
2 Ultimate load analysis of collapse load behaviour.

In the next few sections we shall consider only the ultimate load analysis of reinforced concrete. It should be pointed out that much structural design is now being based on the ultimate loads. An example is CP110 *The Structural Use of Concrete* which was published in 1972 by the British Standards Institution. The Code consists of three volumes, Part 1, Part 2 and Part 3. Part 1 covers the design, materials and workmanship of reinforced, prestressed and precast concrete, Parts 2 and 3 consist wholly of design charts for reinforced concrete beams and columns and prestressed concrete beams.

12.2   The Concrete Code: CP110

An approach to the ultimate load analysis of reinforced concrete has been set out in CP110, *The Structural Use of Concrete*. We give here only the simplest outline.

Two elements come together in a structure: the material strengths and the loads. To obtain the material strengths (called by CP110 the 'characteristic strengths'), tests are carried out on a number of specimens. For concrete, 100 cubes are loaded until failure and the failure stresses are recorded. The cube strength of the concrete f_{cu} is then taken to be that strength below which no more than 5 per cent of the results fall. The yield stress f_y for steel is similarly defined. These stresses f_{cu} and f_y are further reduced to allow for the possible differences between test specimens and the material of the actual structure. The resulting design strengths are $0.4\,f_{cu}$ and $0.67\,f_y$ (columns) and $0.87\,f_y$ (beams).

The ultimate loads for the structure are the working loads multiplied by a factor of safety. The safety factors are 1.4 (dead) and 1.6 (live). To make matters simpler for this chapter we will take a mean value of 1.5.

Let us summarise:

Ultimate stress in concrete is $0.4\,f_{cu}$ for columns and beams
Ultimate stress in steel is $0.67\,f_y$ for columns $0.87\,f_y$ for beams

Popular values are $\qquad f_{cu} = 30\text{ N/mm}^2$
$$f_y = 250\text{ N/mm}^2 \text{ (mild steel)}$$
$$f_y = 425\text{ N/mm}^2 \text{ (high yield steel)}$$

Finally,

ultimate load = $1.5 \times$ working load.

12.3   The ultimate load analysis of the column

The key to ultimate load analysis is to assume that at the section being considered, both the steel and the concrete realise their full ultimate stresses. So for a column the concrete stress is assumed to be $0.4\,f_{cu}$ and the steel stress $0.67\,f_y$. Using the concrete area, A_c, and the steel area, A_s, gives the ultimate load, N.

Clearly,

$$N = 0.4\,f_{cu}A_c + 0.67\,f_yA_s$$

which can be found in CP110, section 3.5.3.

Example 1 (Fig. 12.1)

Calculate the ultimate and working loads for this column using (*a*) mild steel (*b*) high-yield steel. For the concrete take $f_{cu} = 30$ N/mm².

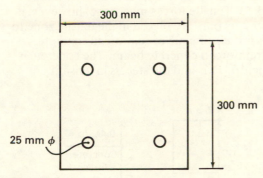

300 mm

300 mm

25 mm ϕ

Fig. 12.1

Simple geometry gives us

area of steel, $A_s = 1963$ mm²
area of concrete, $A_c = 88\,037$ mm²

Ultimate steel stress = $0.67\,f_y = 167$ N/mm² (mild steel) and 285 N/mm² (high yield steel)

Ultimate concrete stress = $0.4\,f_{cu} = 12$ N/mm²

Hence, using load = stress × area, we obtain for the ultimate loads (N),

Mild steel: $N = (12 \times 88\,037 + 167 \times 1963)/10^3$ kN
             $= 1056 + 328$
             $= 1384$ kN

and the working load, $W = \dfrac{1384}{1.5} = 923$ kN

And for the high-yield steel:

$N = (12 \times 88\,037 + 285 \times 1963)/10^3$ kN
  $= 1056 + 559$
  $= 1605$ kN

and the working load, $W = \dfrac{1605}{1.5} = 1070$ kN

It is of interest to note that the steel takes 24 per cent and 35 per cent of the total load in the two cases.

12.4 The ultimate load analysis of the beam

Concrete is weak and unreliable in tension. So much so that in analysis no use is made of the tensile strength of the concrete at all – all tensile zones are assumed to be cracked. Tensile forces must arise, however, if a beam is to take a moment so steel bars are placed in the tensile zone to take these tensile forces. This is the principle of reinforced concrete. Figure 12.2a shows a typical reinforced concrete beam. The concrete is in compression down to the neutral axis and the steel is in tension.

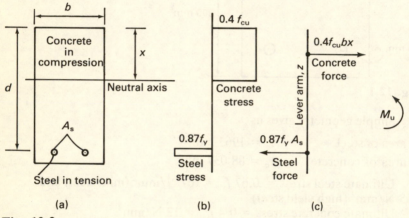

Fig. 12.2

In order to calculate the ultimate moment, M_u, that this section can take, we assume all the concrete above the neutral axis is stressed to its ultimate stress of $0.4\,f_{cu}$ and the steel reinforcement is stressed to $0.87\,f_y$. These stresses are shown in Fig. 12.2b.

The forces from these stresses are easily found by multiplying by the areas bx and A_s for the concrete and steel respectively.

Clearly,

concrete force = $0.4\,f_{cu}bx$

and   steel force = $0.87\,f_y A_s$

These two forces, spaced apart by the lever arm (z), are in equilibrium with the ultimate moment M_u which causes them (Fig. 12.2c). Resolving and taking moments will give all we need to know. We solve first for the depth of the neutral axis (x).

Resolving horizontally.

concrete force = steel force

So,

$$0.4 f_{cu} \, bx = 0.87 f_y A_s$$

Hence,

$$x = (0.87 f_y A_s)/(0.4 f_{cu} b)$$

From Figs. 12.2a, c we obtain,

lever arm, $z = d - \dfrac{x}{2}$

Hence, taking moments about the concrete force,

M_u = steel force × lever arm

$ = 0.87 f_y A_s z$

Example 2 (Fig 12.3)

Calculate the ultimate moment of resistance of this beam. Take $f_{cu} = 30 \text{ N/mm}^2$ and $f_y = 425 \text{ N/mm}^2$. What load can it support over a 6 m span?

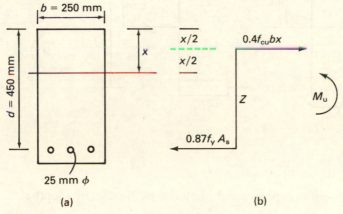

Fig. 12.3

Referring to Fig. 12.3b, and resolving horizontally,

concrete force = steel force

$$0.4 \times 30 \times 250 \, x = 0.87 \times 425 \times 1473$$

Hence,

$$x = 181 \text{ mm}$$

208

So,

$$z = 450 - \frac{181}{2} = 360 \text{ mm}$$

Taking moments about the concrete force,

$$M_u = 0.87 \times 425 \times 1473 \times 360/10^6 \text{ kN m}$$
$$= 196 \text{ kN m}$$

This beam can support w kN/m over 6 m span, where

$$1.5 \times \frac{w \times 6^2}{8} = 196$$

So,

$$w = 30.0 \text{ kN/m}$$

Example 3 (Fig. 12.4a)

Construct a design chart to determine the ultimate moment of a given reinforced concrete beam.

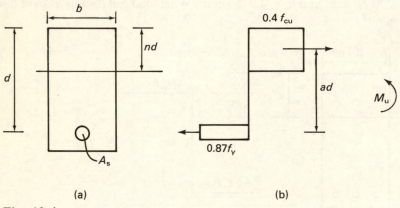

(a)   (b)

Fig. 12.4

It is convenient to express x and z in terms of the effective depth d.

So we take $x = nd$, $z = ad$ (Fig. 12.4b)

Resolving horizontally, $0.87 f_y A_s = 0.4 f_{cu} bnd$

Hence,

$$100\left(\frac{A_s}{bd}\right) = 100\left(\frac{0.4 f_{cu}}{0.87 f_y}\right) n \qquad (1)$$

Taking moments about the steel force,

$$M_u = 0.4 f_{cu} bnd \cdot ad$$

Hence,

$$\frac{M_u}{bd^2} = 0.4 f_{cu} na \qquad (2)$$

Finally,

since $\quad z = d - x/2$

we have $\quad a = 1 - n/2 \qquad (3)$

Using equations (1), (2), (3), we may put $n = 0.1, 0.2, 0.3, 0.4, 0.5$ (with $f_{cu} = 30$ N/mm², $f_y = 425$ N/mm²) to obtain Table 12.1

Table 12.1

M/bd^2	1.14	2.16	3.06	3.84	4.50
$100\,(A_s/bd)$	0.324	0.648	0.973	1.298	1.622

These figures are plotted in Fig. 12.5 and are also given in CP110, Part 2, Chart 3.

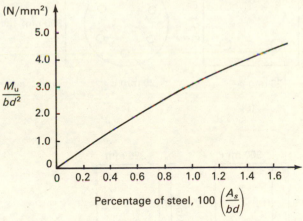

Fig. 12.5 Design chart for reinforced concrete beam

Example 4 (Fig. 12.3)

Use the design chart to obtain M_u.

Firstly,

$$100\left(\frac{A_s}{bd}\right) = 100\left(\frac{1473}{250 \times 450}\right) = 1.3 \text{ per cent steel.}$$

210

So from the graph,

$$\frac{M_u}{bd^2} = 3.9$$

Hence,

$$M_u = 3.9 \times 250 \times 450^2/10^6 \text{ kN m}$$
$$= 197 \text{ kN m as before.}$$

12.5 Exercises

Take $f_{cu} = 30$ N/mm^2, $f_y = 425$ N/mm^2 unless stated otherwise.

1 Distinguish between 'elastic' and 'ultimate' behaviour of a structure.
2 Discuss briefly how reinforced concrete overcomes the weakness of concrete in tension.
3 Explain the symbols f_{cu} and f_y.
4 Calculate the ultimate load that can be supported by the columns in Fig. 12.6.

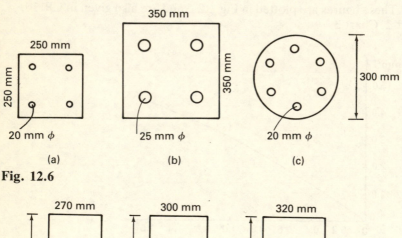

Fig. 12.6

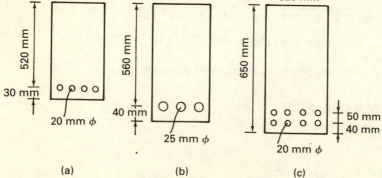

Fig. 12.7

5 Calculate the ultimate moment that can be taken by the beam sections shown in Fig. 12.7.

6 Plot the design chart of Fig. 12.5 to a larger scale.

7 Plot a design chart similar to Fig. 12.5 but based on $f_{cu} = 30$ N/mm^2 and $f_y = 250$ N/mm^2.

8 Devise a design chart for columns. *Hint:* Divide the equation $N = 0.4f_{cu}A_c + 0.67 f_y A_s$ by bd and put $A_c = bd - A_s$. Now plot (N/bd) against $100 (A_s/bd)$.

9 Answer Question 4 using your design chart for columns (see Question 7).

10 Answer Question 5 using your design chart for beams (see Question 8).

Chapter 13

Prestressed concrete

13.1 The principle behind prestressed concrete

Concrete, it is well known, is weak in tension. So if it is possible to avoid tensile stresses occurring in the concrete this will be all to the good. It is not at all obvious how this may be done with a beam as it would appear that under any normal loading the lower half of the beam would be in tension. Reinforced concrete, of course, puts steel in the concrete to take the tensile forces but the idea behind prestressed concrete is different. The simple but brilliant idea is to compress the beam along its length. Since every part of the beam is in compression this means that some load may be supported by the beam without any tension occurring. The application of the load will reduce the compression in the lower part of the beam but it will not produce tension.

Look at Fig. 13.1a. Here a normal beam with its loading has a central bending moment M giving rise to stresses M/Z in the top and bottom edges. Now suppose (Fig. 13.1b) the beam is compressed by an axial force H. An additional compressive stress of H/A is added to the M/Z stresses. If H is chosen so that $H/A = M/Z$ then in fact the tension in the lower edge is reduced to zero – at the centre section at least.

The usefulness of the beam may be further increased by lowering the prestressing force below the neutral axis. In this case what we are trying to do is to arch the beam upwards as much as possible with the prestress until the top edge has tension just about to appear (that is,

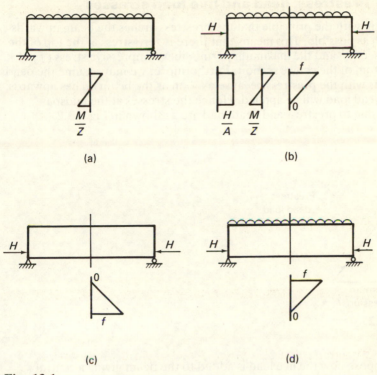

(a)

(b)

(c)

(d)

Fig. 13.1

zero stress). This shown in Fig. 13.1c. The addition of the load will simply reverse this diagram, as shown in Fig. 13.1d.

The analysis of a prestressed beam is largely concerned with the analysis of stress diagrams such as these and in the next section we turn to a more detailed look at them.

Finally, a note about the word 'prestress'. The 'pre' means before the live load comes on to the beam, that is the beam is stressed before the loads are applied. Two other words that need to be understood are 'pre-tensioned' and 'post-tensioned'. Here these words apply to the tensioning of the prestressing wires before ('pre') and after ('post') the concrete has set. The pre-tensioned prestressed beam is usually mass-produced in a factory and its analysis does not concern us here. The post-tensioned prestressed beam is usually made up on site; the beam is constructed with a duct for the prestressing wires and, when the concrete has hardened, so the wires are tensioned, anchored to the end of the beam and released to put the beam in compression. This chapter is concerned only with the post-tensioned beam and its elastic analysis.

13.2 Prestress, dead and live load stresses

We begin with the principle that the prestress arches the beam upwards
as much as possible. This means that there is zero stress at the top of the
centre section and the maximum permissible compressive stress (f) at
the bottom of the centre section. Furthermore, we may assume the dead
load acts with the prestress because as soon as the beam arches upwards
so the dead load will be applied. Hence the stresses at the midspan
section due to prestress and dead load are as shown in Fig. 13.2.

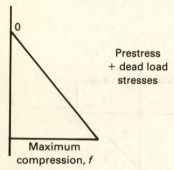

Prestress
+ dead load
stresses

Maximum
compression, f

Fig. 13.2

Suppose now the live load is added to the beam giving a central
moment of M_L. Then bending stresses of M_L/Z will be added to the
existing prestress and dead load stresses to give the resulting stress
shown in Fig. 13.3. With the live load the situation is now reversed – the
beam is bent downwards as far as possible, yet still with no tension.

Let us now add up the stresses in the top (Fig. 13.3) to obtain

$$\frac{M_L}{Z} = f$$

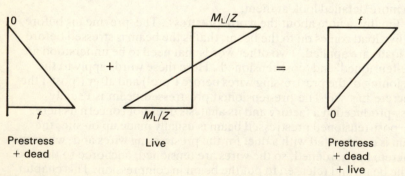

Prestress
+ dead

Live

Prestress
+ dead
+ live

Fig. 13.3

This may be rearranged to give:

Section size for prestressed beam $\quad Z = \dfrac{M_L}{f}$

where M_L is the live load moment and f is the maximum permissible compressive stress in the concrete.

Corollary The size of a prestressed concrete section may be chosen from the live load moment alone – without reference to the dead load.

Example 1

Choose a suitable rectangular section for a prestressed beam to support 28 kN/m over a 12 m span. Allow 18 N/mm^2 compression in the concrete.

We have

$$M_L = \frac{wl^2}{8} = \frac{28 \times 12^2}{8} = 504 \text{ kN m}$$

From Fig. 13.3, equating the top stresses,

$$\frac{M_L}{Z} = f$$

So

$$Z = \frac{M_L}{f}$$

$$= \frac{504 \times 10^6}{18}$$

$$= 28.0 \times 10^6 \text{ mm}^3$$

Putting, $\quad Z = \dfrac{bd^2}{6}$

gives $\quad \dfrac{bd^2}{6} = 28 \times 10^6$

so $\quad d^2 = \dfrac{168 \times 10^6}{b}$

Trying some values of b, we obtain

$b = 300$ m, $\quad d = 750$ mm as a suitable section.

Example 2

Determine the maximum live load that the section shown in Fig. 13.4 can support over a 20 m span. Allow a concrete stress of 16 N/mm^2. Take $I = 10.4 \times 10^9$ mm^4.

216

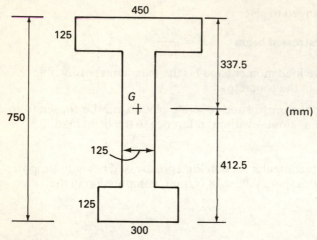

Fig. 13.4

Clearly,

$$Z = \frac{I}{y_{max}} = \frac{10.4 \times 10^9}{412.5} = 25.2 \times 10^6 \text{ mm}^3$$

From Fig. 13.3,

$$\frac{M_L}{Z} = f$$

So,

$$M_L = 16 \times 25.2 \times 10^6/10^6 \text{ kN m} = 403 \text{ kN m}$$

So if the live load is w kN/m, we obtain

$$\frac{w \times 20^2}{8} = 403$$

Hence,

$$w = 8.1 \text{ kN/m}$$

13.3 The prestressing force and the cable eccentricity

We come now to consider what the prestressing force should be and what its eccentricity should be to obtain the best use of the section. The rule is that the dead load acts with the prestress, and the resultant stresses should be those of Fig. 13.2. Now Fig. 13.5 shows the beam

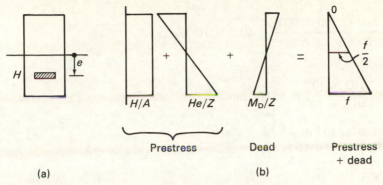

Fig. 13.5

section with the prestressing force (H) eccentric e below the neutral axis.

The stresses due to the prestress are in two parts, H/A and He/Z, as shown in Fig. 13.5b. Note, too, the dead load stresses M_D/Z and the final stresses. Now these stress diagrams will give the information necessary to determine H and e.

How to find the prestressing force and its eccentricity?
Equating centre stresses,

$$\frac{H}{A} = \frac{f}{2} \quad \text{(this will give } H\text{)}.$$

Equating lower stresses,

$$\frac{H}{A} + \frac{He}{Z} - \frac{M_D}{Z} = f \quad \text{(this will give } e\text{)}.$$

Note: Take care of the signs! Why is it $-M_D/Z$? What do you obtain if you equate the top stresses?

Example 3
A rectangluar section 300 mm × 900 mm spans 16 m. Determine the prestressing force and its eccentricity to make best use of the section. Take $f = 17$ N/mm².
The self weight of the beam is

$$w = (2400 \times 9.81/10^3) \times 0.300 \times 0.900$$
$$= 6.36 \text{ kN/m}$$

Hence,

$$M_D = \frac{wL^2}{8} = \frac{6.36 \times 16^2}{8} = 203 \text{ kN m}$$

But,

$$Z = \frac{bd^2}{6} = \frac{300 \times 900^2}{6} = 40.5 \times 10^6 \text{ mm}^3$$

So

$$\frac{M_D}{Z} = \frac{203 \times 10^6}{40.5 \times 10^6} = 5.0 \text{ N/mm}^2$$

Now, referring to Fig. 13.5:
Equating centre stresses,

$$\frac{H}{A} = 8.5$$

so

$$H = 8.5 \times 300 \times 900/10^3 = 2295 \text{ kN}$$

Equating lower stresses (putting $H/A = 8.5$),

$$8.5 + \frac{2295 \times 10^3 \times e}{40.5 \times 10^6} - 5.0 = 17$$

Hence,

$$e = 238 \text{ mm}$$

Q: Where is the prestressing force in relation to the middle third?
A: It is below the middle third by a distance M_D/H. In this case
   $M_D/H = 88$ mm, so $e = 150 + 88 = 238$ mm. *Can you see why?*

It is usual to curve the prestressing cable in a parabolic profile between the support and midspan. There are two reasons for this. Firstly, if the eccentricity were constant then tension would appear at the top of the beam at the support since the dead load moment is zero at the support. To avoid this would mean keeping the cable within the middle third and the full potential of the section would not be realised at midspan. Secondly, the shear forces are reduced if the cable is inclined upwards at the support.

To obtain the offsets which will give a parabolic shape is a simple matter which is illustrated by the next example.

Example 4

A prestressed concrete beam spans 12 m and has its prestressing cable in a parabolic profile with a central eccentricity of 260 mm.
   Calculate the offsets at 1 m centres across the span.

First number the spaces from the centre with the odd numbers 1, 3, 5, . . . (Fig. 13.6). Working back, we may write in the lengths of the offsets beginning with zero and adding in the odd numbers as we go until

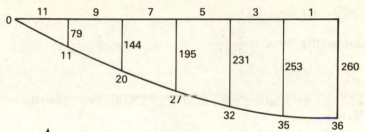

Fig. 13.6

we reach the centre offset (in this case 36). Each offset is now multiplied by 260/36 to obtain the parabola with 260 at the centre. Simple! The values are, 0, 79, 144, 195, 231, 253 and 260 mm.

13.4 Deflection and other factors

Prestressed concrete has very many finer points and only the first few important steps have been outlined.However, we now give a short mention of some additional factors to be considered.

Deflection of a prestressed beam

y = (Downward deflection due to loads)
    − (Upward deflection due to prestress)

The deflection due to the loads is straightforward and has been dealt with in section 6.6. How can we find the upward deflection due to prestress? Suppose the prestressing force H is eccentric e at any section of the beam. Then the bending moment due to prestress is simply $M = He$. In fact e gives the shape of the BM diagram. So we may easily find the bending moment diagram due to prestress which will yield the prestress deflection by the methods of section 6.6.

This may be further simplified for the popular case of the prestressing cable lying in a parabolic curve from the neutral axis at the supports to its maximum eccentricity at midspan. In this case the bending moment diagram is parabolic and so the prestress is equivalent to an upward-acting UDL of w N/mm^2. To obtain the value of w the moment due to w is equated to the moment due to prestress.

So

$$\frac{wL^2}{8} = He$$

Hence,

Upward-acting UDL equivalent to prestress $\quad w = \dfrac{8He}{L^2}$

We quote again from section 6.6:

Deflection due to UDL $\quad y = \dfrac{5wL^4}{384EI}$

Example 5

The beam of Example 3 carries a live load of 20 kN/m. Determine the central deflection.

Since $H = 295$ kN, $e = 238$ mm, the equivalent upward acting UDL from the prestress is

$$w = \frac{8He}{L^2} = \frac{8 \times 2295 \times 0.238}{16^2} = 17.1 \text{ N/mm}$$

Taking, $\quad E = 25$ kN/mm^2

and $\quad I = \dfrac{bd^3}{12} = 18.2 \times 10^9$ mm^4

and substituting into

$$y = \frac{5wL^4}{384EI}$$

we obtain (watch the units! – use kN and mm),

load deflection, $\quad y_1 = \dfrac{5 \times 20 \times 10^{-3} \times 16\,000^4}{384 \times 25 \times 18.2 \times 10^9}$

$$= 37.5 \text{ mm (down)}$$

prestress deflection, $\quad y_2 = \dfrac{17.1}{20} \times 37.5 = 32.1$ mm (up).

Hence,

net deflection, $y = 37.5 - 32.1 = 5.4$ mm down

and we note how helpful the prestress is in lowering the deflection. Here it is just under span/2500, whereas for reinforced concrete the deflections are around span/250 – some ten times larger.

Did you follow the units? 20 kN/m is 20×10^{-3} kN/mm. If you do not like these units then choose your own but be consistent.

Another factor in prestressed concrete is – the tensile stresses. Even though there are no longitudinal tensile stresses, there may be tension in some other direction. Near the support the shear stresses will give rise to a tensile principal stress. Mohr's circle will reveal its value. Suppose the longitudinal bending stress is f and the shear stress is q (Fig. 13.7a) then the Mohr's circle (Fig. 13.7b) will give the principal

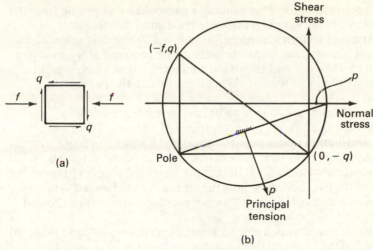

Fig. 13.7

tension, p. Alternatively, it may be calculated from the equation following.

Principal tension in prestressed concrete $\qquad p = \sqrt{\left(\frac{f}{2}\right)^2 + q^2} - \left(\frac{f}{2}\right)$

Other factors are the losses in prestressing force when the force is transferred to the beam and losses due to the creep and shrinkage of the concrete and due to frictional forces in the cable duct. These are usually allowed for by a certain percentage increase in the prestressing force. We need not go into this here but the principles of this chapter will still hold good when losses have to be calculated.

One final interesting problem should be noted: the inversion of the prestressed beam. If the beam is designed for the upright position it will be considerably overstressed if it is inverted – a tension of twice the dead load stresses will appear and the beam will probably collapse. It is, however, possible to design the beam so that it may be inverted and no tension appear. Of course some loss in the live load that can be carried is the price that has to be paid for being able to invert the beam.

13.5 Exercises

1 Write about 100 words explaining the principles that underly prestressed concrete.
2 Choose a suitable rectangular section to support 35 kN/m over a 9 m span allowing 16 N/mm^2 compression in the concrete and no tension. What prestressing force is required? What is the central eccentricity

of the prestressing force? Assuming a parabolic cable profile from the neutral axis at the support, calculate the central deflection.

3 A prestressed section measures 250 mm × 750 mm and spans 12 m. What prestressing force and eccentricity are required at the midspan section to make the best use of the concrete? Allow 18 N/mm² in compression and no tension. What live load can the beam then support?

4 A beam 200 mm × 600 mm × 10 m span is prestressed by a force of 1000 kN acting at a constant eccentricity of 100 mm. Determine the stress distribution under prestress and dead load at 1 m centres across the span and present your answer in the form of clear diagrams. What is the maximum live load UDL that the beam can support? Under the action of the live load show how the stresses have changed – taking the same sections as before. Assume no tensile stresses are allowed at midspan.

5 Calculate the central deflection under (a) prestress and dead load, (b) prestress, dead load and live load, for the beam of Question 4.

6 A prestressed beam of section 300 mm × 800 mm is to span 12 m. By considering the stresses at the centre section, determine the prestressing force H and its eccentricity e to make the best use of the section (a) in the upright position, and (b) when the beam is inverted. Allow 15 N/mm² compression and no tension. Compare the live loads that can be supported by the beam in these two cases.

7 Determine the eccentricities of a parabolic profile at 1 m centres across a 10 m span with a central eccentricity of 180 mm.

Part IV

The computer

Chapter 14

BASIC programming for structures

14.1 The importance of the computer

In this chapter we will try to impart some enthusiasm for the computer. It is quite easy to learn sufficient to make it a helpful tool in structural analysis. It can certainly be exciting, too – there is no moment like the one when your first program works and the results pour out before your eyes.

The computer performs two main functions – and does them very rapidly and very reliably. What may take one man a lifetime might be done in a day by a computer.

Main functions of a computer
1 Calculation.
2 Information storage and retrieval.

Note, however, that it will not do your thinking for you. If you do not know how to solve the problem you can be sure the computer will not know either!

14.2 The main elements of a computer

There is no need to understand the inner workings of a computer in order to use one. Figure 14.1 shows the main elements of a computer. The heart of the system is the central processing unit (CPU): control over the instructions, the data and calculations all happens here. There

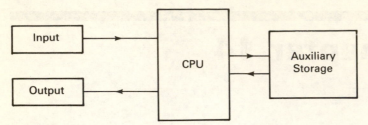

Fig. 14.1

is a certain amount of storage available in the central processor but if
more is required it can access auxiliary storage outside itself.

The input consists of two things: the data and the set of instructions
(called the program) which tells the computer what to do with the data.
The data and program may take different forms, however. They may be
punched on to paper tape or card, or they may be typed directly into a
teletypewriter or VDU (visual display unit – like a television screen).
These machines have simply a typewriter keyboard and paper printout
or screen to show you what you have entered into the computer.

The output device will give you the results. Hopefully the results
will be correct but if not it is likely to be your fault, not the computer's.
The most common form of output is for the results to be printed on to
paper. They may, however, be displayed on the VDU screen or given
on paper tape or card.

The teletypewriter is commonly used as both an input and output
device. It is connected directly to the computer and you can type
your program and data directly into it and receive back the results
immediately. The same applies to the VDU. If you have not sat in front
of one of these machines, do so at the earliest opportunity and try out
one of the programs given later in this chapter.

14.3 The flowchart

Before explaining the program and its language a short word needs to be
said about flowcharts. The flowchart is simply a way of showing the
various steps that are required in the solution of the problem in hand.
The common symbols that are used are shown in Fig. 14.2. and their use
is explained in the examples that follow. The purpose of the flowchart is
to help you hold clearly in your mind the various parts of the problem so
that you have only to put in as much or as little detail as you think
necessary. For many simple problems the flowchart is not necessary at
all and the program may be tackled straight away. For a large problem,
however, the flowchart will repay dividends.

Example 1 (Fig. 14.3)

A simply supported beam carries loads W_1, W_2, W_3, W_4 at distances x_1,
x_2, x_3, x_4, from the left support. Draw up a flow chart for the two reactions.

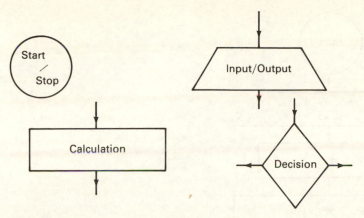

Fig. 14.2

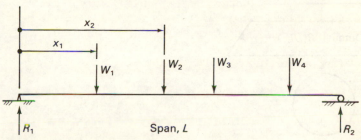

Fig. 14.3

This is fairly straightforward (Fig. 14.4). We just set out the steps that we would normally take – take moments about the left support to give R_2, and resolve vertically to give R_1.

Example 2 (Fig. 14.3)
The beam carries now N loads $W_1, \ldots, W_N$. Draw up a flowchart for the reactions.

This is a little more complicated, at the heart of this is a counter, C, by which the loads can be counted as they are taken in. Look at Fig. 14.5. Note that the statement '$C = C + 1$' simply increases C by 1 Also '$T = T + W$' increases T by W, so T is the running total of the loads. All counters and running totals must first be made equal to zero and that is the reason for $C = 0, T = 0, M = 0$. Now, do you understand this?

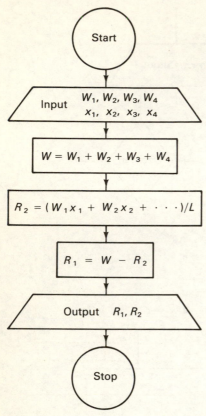

Fig. 14.4

Let us now turn to the most important part of this chapter – the program and its special language.

14.4 The BASIC language

The program has to be written in a definite language and it is important to get the vocabulary and syntax correct. There is very little room for variation – you **must** get it **right!** If you spell 'READ' as 'REED' it just will not work. There are now many different languages, but one of the most popular is BASIC, which is outlined in this chapter. Once you master it you will be able to understand other more powerful languages like FORTRAN with only a little more effort. Here then are the elements of BASIC:

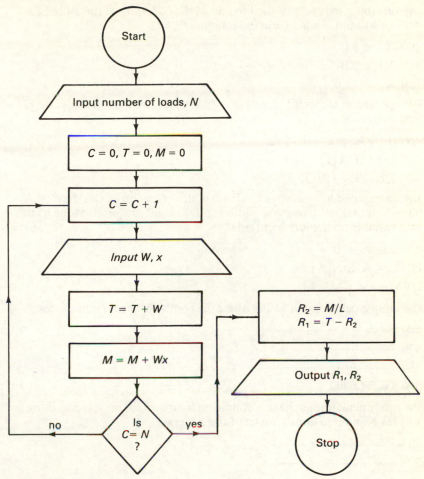

Fig. 14.5

the program: consists of a series of numbered statements which the computer will execute in numerical order.

```
10  INPUT X
20  Y = SQR(X)
30  PRINT Y
40  END
```

the variables: may be assigned to any of the 26 letters of the alphabet with or without a suffix from the integers 0, 1, 2, . . . , 9.

100 X1 = 5.0
110 Y1 = 2.0
120 X2 = 17.0

the operators: are simply, $+ - * / \uparrow$ for plus, minus, multiply, divide and power.

460 T = A $\uparrow$ 2 + B $\uparrow$ 2
470 L = SQR(T)
480 L1 = L $*$ (P/Q)

the statements: have always ONE variable on the left of the sign ' = '. The computer works out the right-hand side and assigns its value to the one variable on the left-hand side.

50 X = A + B + C
60 Y = A/SIN(A)
70 M = M $*$ M $*$ M

the jumps: pass control to any other part of the program you choose.

400 ⎯⎯⎯⎯⎯⎯⎯⎯⎯⎯⎯
420 ⎯⎯⎯⎯⎯⎯⎯⎯⎯⎯⎯
480 ⎯⎯⎯⎯⎯⎯⎯⎯⎯⎯⎯
500 GOTO 400

the conditional jumps: have a condition built into them: IF something is so THEN jump to another part of the program.

20 ⎯⎯⎯⎯⎯⎯⎯⎯⎯⎯⎯
30 ⎯⎯⎯⎯⎯⎯⎯⎯⎯⎯⎯
40 ⎯⎯⎯⎯⎯⎯⎯⎯⎯⎯⎯
50 IF N < 6 THEN 20

the input: is used to take data into the program. It will expect you to give it when it gets to that line.

400 INPUT N
410 INPUT A, B, C
420 INPUT P1, Q1, R1

the output: simply prints out the results that you require. If you want to print out words just put them within inverted commas.

3500 PRINT N
3510 PRINT L, M, X
3520 PRINT 'FORCE = ', F

the loops: enable you to perform the same operation over and over again. The general form of the first line is, FOR N = A TO B STEP C.

40 FOR N = 1 TO 100

50 PRINT N, SQR(N)

60 NEXT N

the arrays: are sets of numbers (Fig. 14.6). An element in the array is specified by a suffix, A(2) = 7, or B(3, 2) = 1. The size of the array must be declared first with the word DIM.

10 DIM A(4)

20 FOR I = 1 TO 4

30 INPUT A(I)

40 NEXT I

$$
A = \begin{bmatrix} 6 \\ 7 \\ 9 \\ 1 \end{bmatrix} \qquad
B = \begin{bmatrix} 2 & 7 & 9 & 4 \\ 3 & 0 & 8 & 6 \\ 0 & 1 & 9 & 2 \end{bmatrix}
$$

Fig. 14.6

subroutines: are used when the same calculation has to be done in a number of different places in the program. The general form is shown in Fig. 14.7 and gives a convenient way of setting them out.

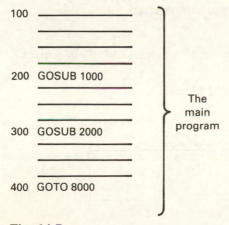

Fig. 14.7

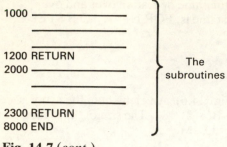

Fig. 14.7 (*cont.*)

There is more than enough here to get you started. At the earliest opportunity try some of it out. As you become more familiar with programming you may want to know more, in which case look out a BASIC manual. For example, a number of standard functions are available in BASIC: SQR(X) or SQRT(X) for $\sqrt{x}$ and other obvious ones like SIN(X), COS (X) and ARCTAN(X). For e^x and $log_e x$ there is EXP(X) and LN(X). Finally, there is the useful function INT(X) which will round X to the nearest integer towards zero, so INT(16.9) = 16 and INT(−3.7) = −3.

14.5 The program

The program is the complete set of instructions required for the solution of the problem. We will assume it is written in BASIC. It may then be typed into the computer using a terminal − either a teletypewriter or VDU. Some command is then required to tell the computer to run your program. This command is usually RUN. The data is typed in as the program requires it and then you just sit back and wait for the results.

Here are a few examples. At the first available opportunity type them into a terminal and see them work for yourself.

Example 3 (Fig. 14.8)

Write a program which will give the bending moment M at 1 m centres across the span of the given beam.

Clearly, $R = 28 \times 18/2 = 252$ kN

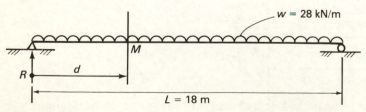

Fig. 14.8

Hence, $M = 252d - 28d(d/2)$
$= 252d - 14d^2$

We want to calculate M for $d = 0, 1, 2, \ldots, 18$ m, so the program looks like this:

```
10  FOR D = 0 TO 18
20  M = 252 * D - 14 * D * D
30  PRINT D, M
40  NEXT D
50  END
```

Simplicity itself. Try it and see. Now the results would look better if there were a heading to them. This is easily done by the addition of the following lines:

```
5  PRINT "BENDING MOMENTS ACROSS BEAM"
6  PRINT "DISTANCE", "MOMENT"
7  PRINT "(METRES)", "(KN. M)"
```

Example 4
Extend the previous problem to deal with any span and any loading.

We have simply to work in terms of the loading w kN/m and the span L metres.

Clearly, $R = wL/2$
So $M = Rd - wd^2/2$

The program now looks like this:

```
10  PRINT "WHAT SPAN (METRES)?"
20  INPUT L
30  PRINT "WHAT LOADING (KN/M)?"
40  INPUT W
50  PRINT "DISTANCE", "MOMENT"
60  R = W*L/2
70  FOR D = 0 TO L
80  M = R * D - W * D * D/2
90  PRINT D, M
100  NEXT D
1000  END
```

Example 5
Extend the previous problem to deal with a set of beams of different spans and loadings.

234

The general case has been solved so we need only add a few lines to make the computer go back to do another example. Perhaps the best way is to use a loop by adding these lines:

```
 5 PRINT "HOW MANY BEAMS DO YOU HAVE?"
 6 INPUT N
 7 FOR I = 1 TO N
200 NEXT I
```

The program of Example 4 of course fits between lines 7 and 200 – but we may add them at the end and the computer will automatically sort them out.

Example 6 (Fig. 14.9)

Write a program to determine the range of values of W for which this column is safe. The column is the smallest Universal Column and is 3 m high with partial end fixity.

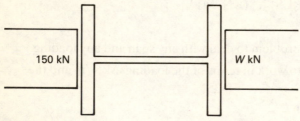

Fig. 14.9

Before writing the program some work needs to be done to evaluate

$$G = \frac{f_c}{p_c} + \frac{f_{bc}}{p_{bc}} \quad \text{in terms of } W.$$

After simplification we obtain this result:

For, $W \leqslant 150$ kN, $G = 1.397 - 3.57 \times 10^{-3} W$
For, $W > 150$ kN, $G = 9.31 \times 10^{-3} W - 0.536$

Let us take values of W from 0–200 kN and write the program as follows:

```
  5 PRINT "LOAD, W", "FACTOR, G"
 10 FOR W = 0 TO 200 STEP 10
 20 IF W > 150 THEN 100
 30 G = 1.397 - 3.57 * W/1000
 40 GOTO 200
100 G = 9.31 * W/1000 - 0.536
```

```
200  PRINT W, G
300  NEXT W
900  END
```

Suppose we wished to print out a third column with SAFE or UNSAFE in it. How could it be done? Something like this must be added:

```
200  IF G > 1 THEN 250
210  PRINT W, G, "SAFE"
220  GOTO 300
250  PRINT W, G, "UNSAFE"
300  NEXT W
```

Quite simple really. Try it out and see.

Example 7 (Fig. 14.10)

A three-pin arch of span 50 m has a central rise of 6 m and carries a central point load of 300 kN. Write a program to give the bending moments in the left of the arch.

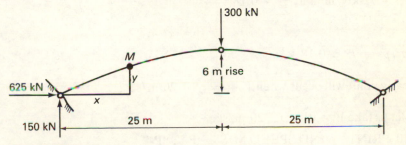

Fig. 14.10

The arch has equation,

$$y = 0.48\,x - 0.0096x^2$$

The BM is

$$M = 150x - 625y$$

So

$$M = 6.00x^2 - 150\,x$$

Taking $x = 0, 5, 10, \ldots, 25$ we obtain the following program,

```
10  PRINT "THREE PIN ARCH MOMENTS"
20  PRINT "X (METRES)", "M (KN. M)"
30  FOR X = 0 TO 25 STEP 5
```

```
40  M = 6.00 * X * X − 150 * X
50  PRINT X, M
60  NEXT X
70  END
```

Example 8

A wall 3 m high and 300 mm thick is subjected to horizontal thrust from wind at V m/sec. Show how the base pressures vary with V.

The base pressures are, $\qquad p_1, p_2 = \dfrac{W}{A} \pm \dfrac{M}{Z}$

where

$A = 0.3$ m², $\qquad Z = 0.015$ m³, $\qquad W = 21.6$ kN

and, taking a wind pressure of $0.6\ V^2$ N/m²,

$M = 0.6\ V^2 \times 3 \times 1.5/1000 = 0.0027\ V^2$ kN m

Hence,

$\dfrac{W}{A} = 72$ kN/m² and $\dfrac{M}{Z} = 0.18\ V^2$

So,

$p_1, p_2 = 72 \pm 0.18\ V^2$

Clearly uplift will occur when $V = \dfrac{72}{0.18} = 20$ m/sec.

So let us take $V = 0, \ \ldots, 20$ m/sec.

```
10  PRINT "WIND SPEED M/S", "P₁", "P₂"
20  FOR V = 0 TO 20
30  P₁ = 72 + 0.18 * V * V
40  P₂ = 72 − 0.18 * V * V
50  PRINT V, P₁, P₂
60  NEXT V
70  END
```

Example 9

Write a program to give the size of a square foundation given the load W and the permissible ground pressure P.

Let the side of the base be D metres.

Then, clearly, $\qquad D = \sqrt{\dfrac{W}{P}}$

The program is therefore like this:

```
10  PRINT "WHAT LOAD?"
20  INPUT W
30  PRINT "WHAT PERMISSIBLE PRESSURE?"
40  INPUT P
50  D = SQR(W/P)
60  PRINT "YOU REQUIRE BASE", D, "METRES SQUARE"
```

Example 10

Write a program to give results from which a design chart for a reinforced concrete beam can be drawn. Take $f_y = 425$ N/mm^2 and $f_{cu} = 30$ N/mm^2.

Using the equations in Chapter 12 we have,

$$\frac{M}{bd^2} = 0.4 f_{cu}\, na = 12\, na$$

and

$$\frac{100A}{bd} = 100\,(0.4\, f_{cu}\, n/f_y) = 2.93\, n$$

So here is the program:

```
  10  PRINT "100A/BD", "M/BD ↑ 2"
  20  FOR N = 0 TO 0.5 STEP 0.1
  30  A = 1 – N/2
  40  Y = 12 * N * A
  50  X = 2.93 * N
  60  PRINT X, Y
  70  NEXT N
1000  END
```

Example 11

Write a program to select a column from the Universal Column tables for a given axial load W kN. Take $p_c = 155$ N/mm^2.

Here we need to store the relevant information from the tables in an array. Let us base the program on the 20 Universal Columns given in Appendix 2.

Clearly, the required area, $\quad A = \dfrac{1000W}{155} = 6.45\ W$ mm^2

$$= 0.0645\ W\ \text{cm}^2$$

The suitable column needs to have a greater area than this.

The program needs to follow these steps:

1 Input the column areas (smallest first).
2 Input load and calculate required area.
3 Select the column and print out the result.

Here we go:

```
 10  DIM C (20)
 20  FOR N = 1 TO 20
 30  INPUT C (N)
 40  NEXT N
 50  PRINT "WHAT LOAD?"
 60  INPUT W
 70  A = 0.0645 W
100  FOR N = 1 TO 20
110  IF A > C (N) THEN 200
120  PRINT "YOU REQUIRE COLUMN NUMBER", N
130  PRINT "OF AREA", C (N), "CM ↑ 2"
140  GOTO 1000
200  NEXT N
300  PRINT "SORRY, NO COLUMN LARGE ENOUGH"
1000  END
```

14.6 The way to success

A vital element in succeeding with the computer is to use it at the earliest opportunity. Learn some of the BASIC language and try it out. Even if it is only PRINT X – *do it!* Go to the computing department and say, 'Excuse me, is there a terminal that I can use with BASIC?' They can only say, 'No'. More than likely they will say, 'Certainly, help yourself . . . use the one over there in the corner.' At this point try to look as though you know what you are doing, march over, switch on (it is probably on already) and start typing. You cannot blow it up. At worst it will print an error message. The best advice is to try a program you know will work – hopefully one of the ones that I have given in this chapter.

If you make a mistake in a line just retype the whole line; it will over-write the line with the mistake in it. When the program is finished then type RUN and the computer will do the rest. There are many other commands which you will find useful. To print out a copy of the program the command varies with different operating systems. Try, LIST. If that does not work then ask someone.

Do not worry if you think you do not know enough. Just keep working at it and you will succeed. I wish you well!

14.7 Exercises

1 What is a terminal? Have you seen one?
2 Explain the letters VDU. Have you used one?
3 Name a popular computer language.
4 What is a program?
5 What is a statement of a program?
6 Give an example of a jump statement.
7 Give an example of a loop. Why is this important?
8 Distinguish between a number and an array.
9 A cantilever of span L metres has a uniformly distributed load w kN/m on it. Work out an expression for the bending moment a distance d metres from the free end. Hence write a program to input values of L and w and output the bending moments at centres of 0.5 m. Use the program for a cantilever of 5 m length carrying a load of 12 kN/m.
10 Write a program which will give the required section size ($b \times d$) of a rectangular prestressed concrete beam given the span (L), loading (w) and permissible concrete stress (p).
11 A three-pin arch spans 100 m with a central rise of 10 m. A load of 100 kN acts 25 m from the left support. The arch equation is $y = 0.4x - 0.004x^2$. The equations for the arch moments are, for $x \leqslant 25$, $M = 75x - 125y$; and for $x > 25$, $M = 75x - 125y - 100(x - 25)$. Write a program to calculate the arch moments at 5 m centres and plot a graph of the results.
12 The top part of Kainji Dam (Fig. 10.13) is a wall of width 5 m. Investigate the vertical stresses in the front and back of this wall when water reaches to the top of the dam. At a depth x metres from the top you may assume the stresses are given by the following equations:

upstream stress, $f_1 = 24x - 0.98x^3$
downstream stress, $f_2 = 24x + 0.98x^3$

Write a program to compute these stresses at 0.5 m centres up to $x = 5.0$ m and plot the variations down each face of the dam.
13 A set of N loads W act at points whose coordinates are (X, Y). Write a program to input this information and output the coordinates of the centre of gravity of the loads.
14 Write a program to input the *area* and r_{yy} for all the Universal Columns. Now input the values of p_c given in Table 7.1 for slenderness ratios lying between 0 and 100. Using these figures as a data base, write a program to select a column to support a given axial load. Assume the column has fixed ends.

Appendices

Appendix 1 Universal Beams: Dimensions and Properties

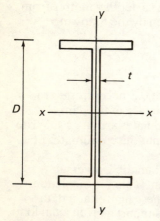

Fig. A1.1

Serial size mm · mm · kg/m	D mm	t mm	Area cm^2	I_{xx} cm^4	Z_{xx} cm^3
914 × 419 × 388	920.5	21.5	494.5	718 742	15 616
914 × 305 × 289	926.6	19.6	368.8	504 594	10 891
838 × 292 × 226	850.9	16.1	288.7	339 747	7 986
762 × 267 × 197	769.6	15.6	250.8	239 894	6 234
686 × 254 × 170	692.9	14.5	216.6	170 147	4 911
610 × 305 × 238	633.0	18.6	303.8	207 571	6 559
610 × 299 × 140	617.0	13.1	178.4	111 844	3 626
533 × 210 × 122	544.6	12.8	155.8	76 207	2 799
457 × 191 × 98	467.4	11.4	125.3	45 717	1 956
457 × 152 × 82	465.1	10.7	104.5	36 215	1 557
406 × 178 × 74	412.8	9.7	95.0	27 329	1 324
406 × 140 × 46	402.3	6.9	59.0	15 647	777.8
356 × 171 × 67	364.0	9.1	85.4	19 522	1 073
356 × 127 × 39	352.8	6.5	49.4	10 087	571.8
305 × 165 × 54	310.9	7.7	68.4	11 710	753.3
305 × 127 × 48	310.4	8.9	60.8	9 504	612.4
305 × 102 × 33	312.7	6.6	40.8	6 487	415.0
254 × 146 × 43	259.6	7.3	55.1	6 558	505.3
254 × 102 × 28	260.4	6.4	36.2	4 008	307.9
203 × 133 × 25	203.2	5.8	32.3	2 356	231.9

Appendix 2   Universal Columns: Dimensions and Properties

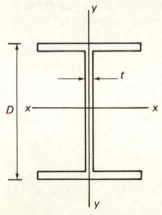

Fig. A2.1

242

Serial size mm . mm . kg/m	D mm	t mm	Area cm²	Z_{xx} cm³	Z_{yy} cm³	r_{yy} cm
356 × 406 × 634	474.7	47.6	808.1	11 592	4 632	11.0
467	436.6	35.9	595.5	8 388	3 293	10.7
340	406.4	26.5	432.7	6 027	2 324	10.4
235	381.0	18.5	299.8	4 153	1 570	10.2
356 × 368 × 202	374.7	16.8	257.9	3 540	1 262	9.57
129	355.6	10.7	164.9	2 264	790.4	9.39
305 × 305 × 283	365.3	26.9	360.4	4 314	1 525	8.25
240	352.6	23.0	305.6	3 641	1 273	8.14
158	327.2	15.7	201.2	2 368	806.3	7.89
118	314.5	11.9	149.8	1 755	587.0	7.75
97	307.8	9.9	123.3	1 442	476.9	7.68
254 × 254 × 167	289.1	19.2	212.4	2 070	740.6	6.79
132	276.4	15.6	167.7	1 622	570.4	6.66
89	260.4	10.5	114.0	1 099	378.9	6.52
73	254.0	8.6	92.9	894.5	305.0	6.46
203 × 203 × 86	222.3	13.0	110.1	851.5	298.7	5.32
60	209.6	9.3	75.8	581.1	199.0	5.19
46	203.2	7.3	58.8	449.2	151.5	5.11
152 × 152 × 37	161.8	8.1	47.4	274.2	91.78	3.87
23	152.4	6.1	29.8	165.7	52.95	3.68

Appendix 3

Proof of $\dfrac{f}{y} = \dfrac{M}{I} = \dfrac{E}{R}$

Proofs have their place in structural analysis – some people think that the place should be as far away as possible – but for my part there is nothing like sitting down to a hot cup of coffee with a good proof to study.

Suppose a moment M bends a unit length of material to a radius R (Fig. A3.1a). Then elementary geometry gives the strain at a distance y from the NA as,

$$\varepsilon = \frac{y}{R}$$

But $\dfrac{f}{\varepsilon} = E$

So $\dfrac{f}{y} = \dfrac{E}{R}$ (1)

Now comes the more difficult part. Look at Fig. A3.1b. Taking

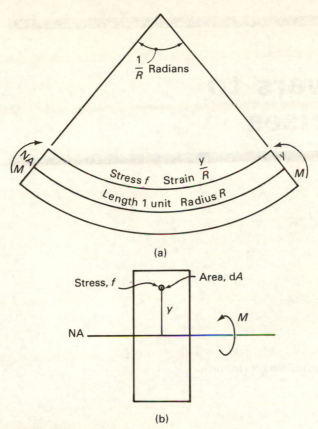

(a)

(b)

Fig. A3.1

moments about the NA gives,

$$M = \int f \mathrm{d}A \cdot y$$

$$= \frac{E}{R} \int y^2 \mathrm{d}A \quad \text{(using Equation 1)}$$

Let us put $\int y^2 \mathrm{d}A = I$ and call it the second moment of area; then

$$\frac{M}{I} = \frac{E}{R} \tag{2}$$

We may now combine Equations 1 and 2 to give,

$$\frac{f}{y} = \frac{M}{I} = \frac{E}{R} \quad \text{which should be learnt by heart.}$$

Finally, putting $\quad Z = \dfrac{I}{y}$

gives $f = \dfrac{M}{Z} \quad$ which should also be learnt by heart.

Answers to exercises

Chapter 1

Q4. 9.81 kN (earth), 1.64 kN (moon).

Chapter 2

Q1. (d), (e), (i) in equilibrium, others not.
Q2. (a) 183.3, 86.7 kN. (b) 231.8, 138.2 kN. (c) 77.3, 82.7 kN.
(d) 314.3, 240.7 kN.
Q3. (a) 25 kN, 0 kN, 75 kN m. (b) 5 kN, 8.66 kN, 30 kN m.
(c) 42.0 kN, 0 kN, 58.8 kN m. (d) 40 kN, 0 kN, 60 kN m.
Q4. (a) 500 kN, 0 kN, 0 kN m. (b) 500 kN, 0 kN, 122.5 kN m.
(c) 300 kN, 0 kN, 70 kN m.
Q5. (a) 312.5, 127.5 kN. (b) −120 kN (down), 320 kN (up).
(c) 300, 200, 0 kN.
(d) $V = 48.6$ kN, $H_1 = 27.3$ kN, $H_2 = 88.6$ kN.
Q6. (a) 137.5, 62.5 kN. (b) 52.1, 47.9 kN.
Q7. (a) 1114 kN, 166.5 kN, 7326 kN m.
Q8. (a) 138.9 kN at 30.26° to horizontal. (b) 38.4 kN at 86.80° to
horizontal. (c) 1200 N halfway up. (d) 18.86 kN at 60.05° to
horizontal.
Q9. (a) 2.286 m from left load. (b) 2.543 m from left load.
(c) 4.087 m from left load. (d) 7.265 m from left load.
Q10. 10.23 m

Q11. (16.51, 5.25) m.
Q12. (a) 285.3 mm.   (b) 250.0 mm.   (c) 322.6 mm.
(d) 270.5 mm.

Chapter 3

Q1. (a) 0, 90 kN, 270 kN m.   (b) 0, −52.2 kN, 208.9 kN m.
(c) 0, 132 kN, 396 kN m.   (d) 0, 40 kN, −80 kN m.
(e) 500 kN, 20 kN, 60 kN m.
(f) 2731 kN, −15.6 kN, 547 kN m.
(g) 47.1 kN, 1.4 kN, −45.0 kN m.
Q2. (a) 69.3 kN, 27.7 kN, −83.1 kN.
(b) 240 kN, −30 kN, −240 kN.

Chapter 4

Q2. (a) 5000 mm^2, 4.17 × 10^6 mm^4, 83.8 × 10^3 mm^3.
(b) 5000 mm^2, 1.04 × 10^6 mm^4, 41.7 × 10^3 mm^3.
(c) 40 × 10^3 mm^2, 133 × 10^6 mm^4, 1.33 × 10^6 mm^3.
(d) 180 × 10^3 mm^2, 5.40 × 10^9 mm^4, 18.0 × 10^6 mm^3.
(e) 187.5 × 10^3 mm^2, 8.79 × 10^9 mm^4, 23.4 × 10^6 mm^3
(f) 187.5 × 10^3 mm^2, 977 × 10^6 mm^4, 7.81.10^6 mm^3.
Q3. (a) 1.59 × 10^9 mm^4, 5.32 × 10^6 mm^3.
(b) 1.49 × 10^9 mm^4, 4.97 × 10^6 mm^3.
(c) 269 × 10^6 mm^4, 1.58 × 10^6 mm^3.
Q4. (a) 12.3 × 10^6 mm^4, 94.8 × 10^3 mm^3.
(b) 632 × 10^6 mm^4, 1.90 × 10^6 mm^3.
(c) 566 × 10^6 mm^4, 1.84 × 10^6 mm^3.
Q5. (a) 100 mm from axis.   (b) 500 mm from axis   (c) 106.2 mm from axis.
Q6. 100.8 mm (above centroid), 123.1 mm (below), 26.4 mm (either side).
Q7. 641 mm (above centroid), 517 mm (below), 595 mm (either side).

Chapter 5

Q3. 97.3 N/mm^2.
Q4. UC 203 × 203 × 86 kg/m.
Q5. 461.9 kN.
Q6. 195 kN/m^2.
Q7. 2.774 m.
Q9. 5.33 kN m.
Q10. 71.1 N/mm^2.

Q11. UB $457 \times 152 \times 82$ kg/m.
Q12. 38.3 kN m.
Q13. 2576.6 kN m.
Q14. 68.8 N/mm^2 (top of web), 109.3 N/mm^2 (neutral axis).
Q15. 60.0 N/mm^2 (top of web), 78.0 N/mm^2 (neutral axis).
Q16. 170 kN/m^2, 80 kN/m^2; 166.7 kN m.
Q17. 3.33 N/mm^2 (comp.), -1.11 N/mm^2 (tens.)
Q18. 111.9 N/mm^2 (comp.), -58.2 N/mm^2 (tens.)
Q19. 121 N/mm^2 (tens.), -21 N/mm^2 (comp.) on planes at 67°, 23° to the horizontal respectively.
Q20. 4 N/mm^2 on a plane at 10° to the horizontal.
Q21. 0.5 N/mm^2 (tension) on plane at 72° to the horizontal.
Q25. 713 N/mm^2; 3.57×10^{-3}, 71 mm.
Q26. 3.25×10^{-3}, 3.25 m.
Q27. (a) 12 N/mm^2, 480×10^{-6}, 813 kN.
(b) 96 N/mm^2, 480×10^{-6}, 283 kN. Permissible load 1096 kN.

Chapter 6

Q6. (a) Reactions 90 kN, 110 kN. Max. BM 280 kN m.
(b) Reactions 381 kN, 309 kN. Max. BM 1602 kN m at 6.17 m from A.
(c) Reactions 295 kN, 275 kN. Max. BM 330.5 kN m at 3.80 m from A.
Q7. (a) 259 kN m, $Z = 1570$ cm^3. (b) 99.6 kN m, $Z = 603$ cm^3.
(c) 224 kN m, $Z = 1359$ cm^3.
Q8. 6.38 kN/m^2.
Q9. 120, 187.5, 270, 337.5, 360 kN m. $Z = 2182$ cm^3.
Q10. (a) Reactions 30 kN, 72 kN m.
(b) Reactions 59 kN, 97.3 kN m.
(c) Reactions 60.8 kN, 126.4 kN m.
Q13. 8.3 mm.
Q14. 240 N/m, 30 000 mm^3.
Q15. 34.6 mm.
Q16. 0.07, 0.23, 0.45 mm.

Chapter 7

Q11. UC $203 \times 203 \times 60$ kg/m.
Q13. $f_c = 133$ N/mm^2, $p_c = 133$ N/mm^2.
Q14. 2837 kN.
Q16. (a) yes. (b) no.
Q17. (a) yes. (b) no.
Q18. (a) unsafe. (b) safe.
(c) safe; 157.4, 93.8, -14.2, -77.8 N/mm^2.

Q19. $W = 0$, $G = 1.314$; $W = 800$, $G = 0.920$; $W = 1000$,
$G = 1.249$. Column safe for $630 < W < 850$ kN.
Q20. 31, 142 N/mm², 987 kN.

Chapter 8

Q1. Lower boom, (a) 0, 65, 130, 155, 105, 70, 35, 0 kN.
(b) 25, 25, 75, 75, 105, 105, 55, 55 kN.
Top boom, (c) 0, 60, 120, 180, 240, 200, 100, 0 kN.
Q2. (a) 80 kN, 42.4 kN. (b) 92.4 kN, 28.9 kN.
(c) 134 kN (comp.), 20 kN (comp.).
Q3. Lower boom, (a) 0, 80, 100, 0 kN. (b) 46, 115, 127, 58 kN.
(c) 67, 67, 67 kN.
Q4. 28 mm.
Q5. 404 kN, 115 kN.

Chapter 9

Q6. 350 kN, 444 kN; bending moments zero.
Q7. −225 kN m, 121 kN.
Q8. $H = 12\,500/r$ kN.

Q9. $H = \dfrac{wl^2}{8r}$

Q10. 140 kN, 60 kN, 150 kN. $M = 0$, 196, 440, 840, 360, 0 kN m.
$N = 204.6$, 201.8, 195.8, 185.5/124.6, 138.6, 150.0 kN.
Q11. $N = 320.3$ kN, $Q = -87.4$ kN.
Q12. $V_1 = 835.6$ kN, $V_2 = 764.4$ kN, $H_1 = H_2 = 675.6$ kN.
Q13. $V_1 = 35.5$ kN, $V_2 = 44.5$ kN, $H_1 = H_2 = 68.2$ kN.
$M = 0$, −1.5, −1.9, −1.1, 0, 0.8, 1.9, 2.2, 1.5, 0 kN m.
Q14. $Q = 51.4$, 90.8/−20.7, 15.2/−92.6, −56.7, −18.6, 17.5,
48.6 kN at 10 m centres across the arch.
Q15. 1875 kN m, −1875 kN m at quarter points.

Chapter 10

Q16. 47.1 kN/m².
Q17. 90.8, 26.8 kN/m², 5.5.
Q18. 1.44 m, 2.08 m.
Q19. $\lambda_s = \infty$, 3.31, 0.83, 0.37; $\lambda_t = \infty$, 35.7, 4.47, 1.32.
Q20. 85.5, 78.3 kN/m², 6.26, 2.67.

Chapter 11

Q9. (a) 187.3, (b) 318.9, (c) 208.0 kN/m².
Q10. 2.45, 3.16, 4.08 m.

Q12.   1089, 5040, 4725 kN.
Q13.   (a) 175, 225;   (b) 174, 274;   (c) 102, 145 kN/m².
Q14.   650 kN m.
Q15.   7.0 m.
Q16.   138.9, 83.3, 27.7, 83.3 kN/m². Centre of foundation at
   (7.33, 8.67) m.
Q17.   parallel sides 1.710 m, 3.770 m distant 4.200 m apart.
Q18.   7.44 m, 301, 259 kN/m².
Q19.   880, 294, 176 m².
Q20.   1.35 N/mm², 199 kN/m²; 212, 186 kN/m².

Chapter 12

Q4.   (a) 1093,   (b) 2005,   (c) 1362 kN.
Q5.   (a) 208,   (b) 264,   (c) 431 kN m.

Chapter 13

Q2.   250 mm × 730 mm, 1460 kN, 152 mm, 7.3 mm.
Q3.   1688 kN, 172 mm, 23.4 kN/m.
Q4.   Support 0, 16.7, midspan 2.9, 13.8 N/mm². Live load,
   13.2 kN/m.
Q5.   9.7 mm up, 9.4 mm down.
Q6.   (a) 1800 kN, 190 mm   (b) 1800 kN, 76 mm. Live loads,
   26.7, 15.3 kN/m.
Q7.   0, 65, 115, 151, 173, 180 mm.

Index